THE WORLD BENEATH THE SEA

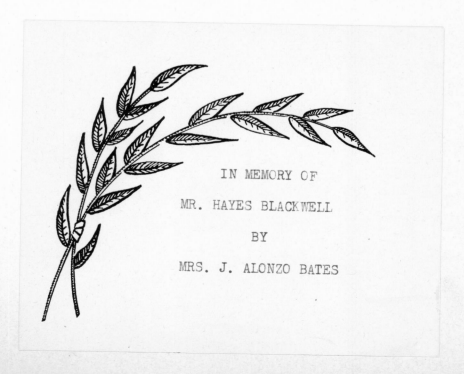

THE WORLD OF NATURE

THE WORLD BENEATH THE SEA

Adapted from the Italian of
MENICO TORCHIO
Director of the Aquarium and
the Hydrobiological Laboratory of Milan

Edited by **Maurice Burton DSc, FZS**
Formerly Deputy Keeper of Zoology,
British Museum (Natural History)

BOUNTY BOOKS
A Division of Crown Publishers, Inc., New York

Contents

Endpapers: Octopus grappling with a lobster (Nilsson/Camera Press)
Title Page: Nurse shark (Popperfoto)

Copyright © MCMLXXII by Istituto Geografico de Agostini, Novara
English edition © MCMLXXII and MCMLXXIII by Orbis Publishing Limited, London
Revised edition © MCMLXXIV by Orbis Publishing Limited, London
Library of Congress Catalog card number: 73-82260
All rights reserved
This edition is published by Bounty Books
a division of Crown Publishing, Inc.
by arrangement with Orbis Publishing Limited
a b c d e f g h
Printed in Italy by IGDA, Novara
Adapted from the original Italian *La Vita nel Mare*

Foreword

by Maurice Burton DSc, FZS
Formerly Deputy Keeper of Zoology, British Museum (Natural History)

It is little more than a hundred years since the study of the seashore first became popular. Before that it was almost the exclusive preserve of a relatively few dedicated naturalists. It is less than a hundred years since marine biology was set firmly on its feet by the establishment in a few countries of marine laboratories and by frequent ocean-going expeditions. Since then, the number of survey ships has increased out of all belief, and marine laboratories dot the coastlines of most of the countries in the world. Above all, the presentation of films of seashore or undersea life, in addition to the many popular books on the subject, has heightened an interest that has grown steadily since the beginning of the century.

The interest has not always been without ulterior motive. With intensification of submarine warfare in both World Wars, hostile navies were compelled to think more about what lay beneath the surface of the sea. Then came a phase in human history dominated by the menace of a human population explosion. Inevitably, people began to look to the sea as a source of extra food. More recently, concern has been expressed about the increasing pollution of the seas.

The present volume deals but passingly with these momentous problems, but not because they are unimportant. Professor Torchio has wisely chosen to deal with the seas, not as a separate entity, but as a part of the global environment: he takes us from the land fringing the margins of the sea, just above high-tide mark, to the shore, where land and marine plants and animals to some extent intermingle. From there he takes us into the shallow seas, the home of shoals of food fishes and of the fascinating reefs, full of colour and animation, which form the undersea gardens, to the ocean abysses. In these regions the eternal darkness is broken only by the living lights carried by the bizarre fishes, prawns and squids. The author portrays the seas in the condition in which they have existed for countless ages, and in the state in which, it is hoped, they will yet continue. The text is wide-ranging and is supported by excellent pictures.

Index

Page references to photographs are printed in *italic* type. Common names are used whenever possible.

Life and the sea

It is known from the study of fossils that the oldest forms of life originated in the primordial seas, and there are many reasons why this should have been so. The sea is extremely rich in nutrients, and, in comparison to other areas of the Earth's surface, it is better protected against cosmic radiations and is more constant and uniform in its physical, chemical, climatic, and biological conditions. Furthermore, in terms of its chemical and physical nature, sea water is rather similar to the internal liquids of many of the animals that populate it today, providing them in effect with an external 'blood' or 'lymph'. In addition, the sea, by virtue of its high density, is well suited to supporting the bodies of living organisms, from the smallest to the largest, from those with a low water content to those composed largely of water, including species that float and those that swim. Thus the sea provides physical support for both the 'lightest' and 'heaviest' in its waters.

Because of complex factors such as these, and many others, the sea has had in earlier geological eras – and still has to some extent – a monopoly of some of the important forms of life. Extinct species, such as the animals called trilobites, and the red alga *Solenopora*, have been shown by fossil evidence to be exclusively marine. Among present-day animals, all the echinoderms and the tunicates, and among plants, all the seaweeds, are found only in marine or at least strongly brackish water.

The blood plasma of present-day man, as indeed of the other living land mammals and of birds, has a salt composition akin to that of the sea at the dawn of the Mesozoic era, that is, the period when the higher animals started to colonize dry land. Initially this animal invasion was limited to those sheets of water closely linked to the newly emerged land, such as the intertidal waters, lagoons, coastal saltings and estuaries. From these it spread out, ultimately to populate all the areas above sea level, from the summits of the highest mountains to the depths of the deepest caves or caverns, from the heart of the driest deserts to the longest rivers and deepest lakes.

Life, having moved into the fresh waters and on to the dry land, encountered a great variety of environmental conditions which differed substantially from those existing in the seas. As a result of the widescale adaptation that subsequently occurred, there developed in the fresh waters and on the dry land a variety of forms infinitely greater than anything witnessed at any stage in the history of the sea.

Today, therefore, the marine species constitute only a small minority of animal and plant species living on our planet. In addition, some marine groups are extremely old, virtually constituting 'living fossils', and many have evolved only slightly since they first became established.

The traffic between sea and dry land has not, however, been completely one way. Some marine groups of animals and plants are derived from dry land forms: for example the cetaceans (whales, dolphins, and porpoises) from the mammals, and the *Elobia* from the monocotyledonous (flowering) plants. These organisms, by moving from the land to the seas, have thus travelled in the opposite direction from most living organisms.

Life in the sea is shaped by many environmental factors, some of which are present in the fresh waters or on dry land, while others are peculiar to marine waters.

These factors, termed hydrographic parameters, influence all aspects of sea life, particularly the distribution of marine organisms. The parameters of the marine environment may be broadly classified as follows: physical (illumination, temperature, pressure, density of the water); chemical (salinity, chlorinity, oxygen content, etc.); chemico-physical (for example,

1

the hydrogen ion concentration); dynamic (wave movements, tides, and currents); and biological.

Before entering into a fuller discussion of these environmental factors and their role in marine life, we shall examine some aspects of the geographical relationship between land and sea, distinguishing between the seas and the other areas of the Earth occupied by water.

The distribution of the seas and dry land

From a cursory glance at a map of the world it is apparent that the land areas are not equally distributed between the Northern and Southern Hemispheres. Yet despite the fact that more than two-thirds of the land surface lies north of the equator, the Northern Hemisphere still shows a greater area of sea than of land. This is easier to appreciate if we remember the global dominance of sea water over land – over seventy per cent of the Earth's surface is covered by sea, accounting for a total area of 139 million square miles. Furthermore, it is interesting to note that almost all the areas of dry land have open waters at their antipodes. That is to say, the land directly opposite them (in a line through the centre of the Earth) is submerged beneath seas. This phenomenon is called the 'diametric opposition of the continents and oceans', and its most obvious example is furnished by the two poles: at the North Pole there is a vast extent of water and no land, while the South Pole falls in a land mass.

The terms 'sea' and 'ocean' are often used interchangeably, and in the strict sense there is no difference between them. However, in some contexts their meanings do differ, and the distinction is most evident when it comes to proper names. The word 'seas' or 'the sea' can refer to the total expanse of the world's salt waters, but specific seas such as the Red Sea and North Sea cover a smaller area than an ocean. Their average depth is less, they connect with less extensive and shallower basins, and they lie closer to a continent. Oceans, on the other hand, are the vast expanses that surround, or separate, the continents or groups of continents.

The seas can be divided into different categories: coastal, mediterranean, and inland. Given this categorization, then, to be correct in our nomenclature we ought to refer to the Persian Sea, Sea of Bengal, and so on, rather than calling them the Persian Gulf or the Bay of Bengal. Five oceans are generally recognized: Atlantic, Indian, Pacific, Arctic and Antarctic. There are some, however, who would consider the Arctic as a part of the Atlantic, while others again would divide the Antarctic and share it out between the Atlantic, Indian and Pacific Oceans.

The east coast of South Africa, looking out on to the Indian Ocean, is a series of bays with gently sloping sandy beaches, and a rocky headland at either end of each bay

Salinity of the water

The salty and somewhat bitter taste of sea water is well known to everyone and is due to its content, in solution, of an average of about 35 grams of salts per kilogram. Among these salts the most important are sodium chloride, magnesium chloride, and magnesium sulphate. The salinity can vary quite considerably from sea to sea, but the ratios between the various salts remain substantially the same, except near the mouths of rivers and in areas which receive torrential discharges from the land or are subject to other exceptional conditions. From this it follows that to determine the approximate salinity of a sea, it is enough to measure one of the component salts.

The immediate question raised at this stage is: why do seas vary in salinity? The answer, put briefly, lies in the wide variation in the Earth's climate and geography. The climatological factor is most evident in low latitudes, where low rainfall and rapid evaporation, aided by poor circulation of the water itself, are responsible for the high salinity levels recorded in this belt. Striking examples are provided by the Sargasso area of the North Atlantic and the South Atlantic off Brazil, where surface salinities are approximately 37 grams per kilogram. In higher latitudes, where heavy rainfall, low evaporation and extensive land drainage predominate, salinity is markedly lower, dropping to as little as 7 g/kg. Nowhere is this more apparent than in land-locked areas of water which receive large additions of fresh water from rivers and from the melting of glaciers. Certain areas, however, represent clear departures from the normal range of salinities in land-locked areas in high latitudes, notably the Red Sea, whose surface salinity is over 40 grams per kilogram, and the eastern Mediterranean, where the salinity may reach 39 g/kg. These exceptional levels are attributable to the high temperature and rate of evaporation found in these enclosed areas.

The salinity is of vital importance for living organisms in the sea, both animal and vegetable. While it is true that these are organisms that can live and reproduce in both saline and fresh waters, there are others that not only must live exclusively in the sea, but require precise conditions of salinity, and will perish if these are significantly modified.

Between these extremes lie a host of marine organisms that can withstand differing degrees of salinity change, including those that live in the open ocean and have only a limited tolerance.

By contrast, there are organisms that can withstand great variations in environmental salinity, notably the plant *Dunaliella salina* and the brine shrimp *Artemia salina*. These species are also remarkable for another quality – toleration of extremely high concentrations of salt which enables them to inhabit hypersaline waters.

The brine shrimp, at most 12 millimetres long, is a primitive crustacean found in many parts of the world, from Greenland to Australia, from the Caribbean to central Asia. It swims upside-down, paddling with its eleven pairs of legs, and its form varies according to the salinity of the water in which it is living. In southern Europe it lives in shallow ponds where sea water has evaporated to give a highly concentrated salt solution. It can be found in salt pools in which crystals form round the edges, in a concentration as high as 30 per cent of salt as compared with the 3.5 per cent of normal sea water. That it can be found also in the salty lakes of central Asia may be the result of its having lived there when the sea penetrated that

Above: Part of the rocky coast of the Indian Ocean in Kenya, near Malindi. Small rock pools interconnect with each other and with the sea

Right: Breakers on the rocky coast at Cape Agulhas, in South Africa

far, a long time ago, as with the Lake Baikal seal, of central Asia. It can also be found in the Great Salt Lake of Utah, where its eggs can be collected by the shovelful. On the other hand, the brine shrimp may lay eggs capable of remaining quiescent for years. Normally, this is no more than a means of survival where lakes may dry out temporarily. It does also introduce the possibility of the eggs being carried about by the wind.

Brine shrimp eggs have been dried in a high vacuum so that practically every trace of water has been removed from them, the chemical processes of life being thus brought to a standstill. The eggs can then be cooled to the temperature of liquid air ($-190°C$, $-310°F$) and will hatch successfully when returned to salt water at normal temperature. Moreover, when fully dry some of the eggs have been known to survive being kept for two hours at a temperature of $105°C$ ($221°F$).

Temperature of the water

Another important aspect of the seas is their temperature, which, generally speaking, decreases from the Equator towards the poles and from the surface towards the bottom. Both these phenomena are explained by the fact that a sea's temperature depends upon its exposure to the sun, and this is greater at the Equator and on the surface than at high latitudes and at depths. The surface temperature in certain enclosed seas reaches as much as $35°C$ (in the Red Sea and the Persian Gulf), and falls below zero in polar waters: sea water freezes at about $-2°C$. In ocean depths a layer of water exists with a temperature around freezing point, but ice is not formed because of the high pressure to which this layer is subjected.

The decrease in the temperature of sea water with increasing depth is generally rapid for the first 150 or 300 feet (50 to 100 metres), but from there on it becomes progressively less. In the oceans of certain temperate zones, the temperature in summer ranges from $26°C$ at the surface to about $13°C$ at a depth of a few hundred or a thousand feet, and to about $4°C$ at a depth of 3000 feet (1000 metres); it then falls much more slowly, decreasing by only two degrees over the course of some 10,000 feet (3000 metres).

In the course of the year the temperature varies both at the surface and down to a certain depth; at the surface it fluctuates by only a few degrees in the torrid zones, and by about twelve degrees in the temperate zones. In that part of the Mediterranean known as the Ligurian Sea, for example, the temperature changes at the surface, from $12°C$ in winter to $25°C$ in summer. Near the

The dramatic cliff formation below the town of Bonifacio in Corsica is the result of continuous pounding by the sea against the sedimentary chalk strata of the cliffs. It has also been affected by weathering

Lebanese coast it varies from 17°C to 29°C.

Temperature is a highly important factor in both the geographical and the vertical distribution of marine organisms. Since the daily and annual temperature fluctuations are greater at the surface than at depth, surface animals and plants can withstand greater variations (and generally have a broader geographical distribution) than those at greater depth. Again, some marine organisms thrive at the surface at a given latitude only during certain seasons. For example, there are Atlantic fishes which migrate into the Mediterranean only in summer; and in the summer the Mediterranean is populated by a large number of species of algae and other groups which have more tropical affinities (*Acetabularia*, *Digenea*, etc), and these are replaced in winter by species with temperate affinities (*Ulothrix*, *Bangia*, *Nemalion* etc).

Among animals there are numerous examples of species that have begun to populate a given area, or conversely to abandon it, as a result of a gradual variation of the ambient temperatures. The Mediterranean, for example, has seen in the course of the Quaternary period (the current geological period) the disappearance or movement into deep waters of various elements of its former

Right: Table Bay, South Africa, with Cape Town in the distance, and beyond it Table Mountain, with sand dunes in the foreground

Below: the relentless pounding of breakers on this shingle beach accounts for the total absence of seaweed

surface fauna, apparently because of the rise in temperature. For this reason it provides a home in its deep waters for fewer species than live in the Atlantic at the same latitudes and depths. Trout (*Salmo*), widespread in the inland waters of northern Europe, frequently come down to the North Sea and the Baltic, returning to the rivers to breed, but the *Salmo* living in the inland waters of Mediterranean Europe only rarely move down to the seas, and then only when the surface waters have a temperature not greater

than 18°C. Probably in the past, at a time when the Mediterranean was colder, the trout came down more frequently, just as the northern *Salmo* do today.

Furthermore, in connection with the gradual rise of the ambient temperature over recent decades in certain areas of the north-east Atlantic, a spread northwards has been observed along the Norwegian coasts of algae and animals that were previously confined to lower and more temperate latitudes.

Solar radiation

Another of the physical factors influencing marine life is solar radiation on the water. This is at a maximum at the surface, since light entering the water is both absorbed and scattered. It is also greater at the Equator where the sun's rays strike the water almost at a right-angle, and consequently a lower proportion is reflected away from the water. Sunlight is composed of light of various wavelengths and these are extinguished at different depths. The red and

9

Above: Palm trees growing down to the edge of the sea round the coast of Malikolo, one of the islands in the archipelago of the New Hebrides

Left: Part of the southerly coast of the Hawaiian island of Oahu. The land is comparatively barren, due mainly to the lack of topsoil on the rock

ultra-violet parts of the spectrum are rapidly absorbed. The green and blue rays penetrate more deeply.

The variations in the colour and intensity of light affect the vertical distribution of many plants and limit the depths at which they can survive. Phototropic vegetation is plant life which needs light for the process of photosynthesis so that distribution is generally limited to 130 to 160 feet (40 to 50 metres) in the northern seas, 320 feet (100 metres) off Florida, 420 feet (130 metres) off Capri, and 580 feet (180 metres) near the Balearics. The maximum depths reported at low latitudes with very clear waters are around 650 feet (200 metres).

Solar radiation is also of great importance to animal life, indirectly as well as directly. It largely determines the temperature of the water and also, by affecting the rate of evaporation, the salinity. Many animals flourish exclusively or predominantly in particular conditions of illumination, and some spend their adult lives fixed to a substratum. Their larvae may settle at a certain depth, or on bottoms with a certain gradient, or in caves, or wherever the preferred light conditions obtain. If they settled anywhere else they would not survive. Many of the planktonic (drifting) and the nektonic (swimming)

11

animals make vertical migrations during the course of each 24 hours, inclining towards those depths in which they find satisfactory conditions of illumination, for here will be the phytoplankton on which they, or their prey, feed. This explains, in part, why numerous animals are found at appreciably different depths by day and by night, some descending at dawn to depths of some hundreds or thousands of feet, moving away from the coasts if necessary to do so, and returning the next evening nearer to the surface, frequently coming into inshore coastal waters. Thus light probably controls the diurnal migrations of zoo-plankton, but the value of such migrations is not necessarily only in placing the species in a specific light intensity *per se*, but in keeping them near the phytoplankton on which they feed.

The colour of the surface water is generally greenish or bluish, the exact shade varying with the temperature, the chemical composition of the water and richness in plankton, or by reason of coloured particles in suspension. It can also vary as a result of reflection due to ice, or to transparency effects that allow colours of the bottom to show through. Neither does a sea remain constant in colour; it can change colour in relation to the atmospheric conditions, and

with other phenomena.

The marine environment can be divided into zones on the basis of the illumination of the water. The shallowest zone, which reaches to a depth varying from 65 to 390 feet (20 to 120 metres) according to certain conditions – mainly the latitude and the transparency of the water – is called the euphotic zone (euphotic means good light). In this, all the red rays and some of the blue are absorbed. The vegetation endowed with chlorophyll receives sufficient solar energy here for photosynthesis to occur. The lower limit of the euphotic zone coincides with the 'compensation depth', the greatest depth at which plants are able to manufacture a quantity of food substances equivalent to – and thus compensatory for – the quantity that they consume in order to live. Below the euphotic zone lies the oligophotic zone (oligophotic means little light), which receives only the most penetrating rays of the solar spectrum. This zone extends to a depth of one to two thousand feet (300 to 600 metres), with an average of about 1600 feet (500 metres). In this zone the classic chlorophyll-carrying organisms are lacking, since the light intensity is too low to allow them to manufacture an adequate supply of food. Below the oligophotic zone

The Arctic: a continental ice-mass meets the sea close to two areas of ice-free landscape. Part of an ice floe is visible in the foreground

Pack ice (smaller than ice floes and ice fields), is made up of floating ice fragments

extends the aphotic zone (no light), which is not capable of sustaining any plant life.

As well as diminishing in intensity the light that has succeeded in penetrating varies in wavelength with depth. Therefore algae that live in the depths enjoy light which is both less intense and of a more limited colour range from that which illuminates the plants of the shallow water. Since it is possible for the algae to have, in addition to chlorophyll, pigments of a different nature, it is thought that a relationship exists between the colour of these accessory pigments and that of the light received at the depths in which various groups of algae live. According to this theory of chromatic adaptation, the red algae are able to live without red radiations since their red pigment allows them to exploit the green-blue light which is directly complementary to red.

Density and pressure of the water

The density of sea water varies in inverse proportion to the temperature, but is directly proportional to the salinity. With each ten metres (about 32 feet) of depth, the pressure increases by about one atmosphere, so that at a depth of 1300 feet (400 metres) it is at a pressure of approximately 40 atmospheres. The water compresses very slightly under the enormous pressures of the great depths, and is therefore more dense than it is at the surface.

Movements of the water

Apart from its physical nature, one of the most important features governing marine life is the movement of the sea.

This movement takes three forms: waves, currents, and tides, all of which will be familiar to anyone who has made even the most elementary study of the sea. The extent of their influence on marine life will become evident as each movement is discussed in turn, starting with the waves.

When the wind blows upon the sea, the surface waters become agitated, taking on an oscillatory motion with waves characterized by a period, a length, a speed of propagation, and a height. The oscillatory motion of the waves can survive long after the wind that produced it has died down, and can also propagate itself at long distances from the point of origin, in areas not affected by the wind. As well as these waves which are created by atmospheric movements, there are those that have a different origin: they can, for example, result from coastal earth tremors or from submarine earthquakes.

The movement of a wave in the sea is not really a true transportation of liquid particles, as can easily be observed by watching the position of a float: the liquid particles, like something passively floating, rise and fall, describing a vertical circle, oriented towards the propagation of the wave.

The wave dies out quickly in the vertical direction. The particles of water describe smaller circles, the farther they are from the surface, down to a depth at which the undulatory movement ceases altogether: although theoretically this occurs at an infinite depth, the effect of the waves is no longer perceptible at about 300 feet (less than 100 metres) below the surface.

The wave undergoes substantial modifications as it approaches the coast, since the particles of water can no longer complete their normal path – the bottom restricts them. Consequently their motion is broken down and the waves pick up particles of the bottom with a true transporting action. Running over its base, which strips the bottom, the surface of the water now moves rapidly until the upper part finally is split forward.

The coastal waves are an important environmental factor for marine organisms, in that their movement provokes a continuous renewal of the water around the expanses of algae, sedentary organisms, submerged reef plants, and in the reefs and tidal pools. Moreover, they impede powerful thermal movements and favour the exchanges of gases in the surface water. Since the 'power' of the wave is extremely great, the organisms living on the rocks pounded by it must have an effective system of anchorage and must be exceptionally robust; it is estimated that a force of about 45 kilograms per square centimetre (about 640 lbs per square inch) must be exerted to fracture the stalk of one alga, *Fucus vesiculosus*, which can live in zones battered by large waves. No algae, however, can withstand the heaviest pounding that waves can produce.

The winds not only initiate an undulatory motion, but also contribute to a large extent to the creation of ocean currents. In fact, maps of the ocean surface currents show a general correspondence with those showing the prevailing winds. What happens is that the wind moves the surface water, which is then replaced by other water from neighbouring zones, a process known as compensatory movement. However, it is not only the wind that is involved in producing currents: the differences in density of the water also have a part to play in their production for warm and less dense water moves towards cold and relatively dense water, with compensatory movements in its wake. For example, the relatively 'light' waters of the Atlantic and 'heavy' waters of the Mediterranean meet at the Straits of Gibraltar. Surface currents are created which move towards the Mediterranean; at depth, currents are encountered that move in the opposite direction.

The currents are of great importance to marine animals and plants, because they affect the nature of the water, notably the temperature, density and salinity. They act as means of transport for certain animals and plants, and this is obviously a contributory factor in the geographical distribution of species. They also bring food to sessile forms.

Another important phenomenon is that of the tides, which are periodic variations of the level of the sea, due principally to the gravitational pull exerted on the waters by the moon and sun. In a period of 24 hours and 50 minutes, in almost every locality, the waters twice rise to a maximum level and recede to a minimum level. The size of the tide is the change of level between consecutive high and low tides. This varies greatly from one locality to another, from a few inches to many yards, and in the same locality it varies periodically in time, increasing from one to two days after the first and last quarters of the moon until one to two days after the full or the new moon. This periodical variation is due to the fact that at full and new moon the forces of attraction of sun and moon operate together, while at the quarters they are in opposition.

The tides have a considerable influence upon animals and plants, particularly in seas where the tides have a large amplitude, since the daily exposure of a coastal belt determines the nature of its flora and fauna and the adaptations that enable aquatic organisms to survive the period of exposure. Furthermore, the behaviour of the tides brings about horizontal zoning of the organisms in the littoral (the area between high and low tides) with a well-defined vertical succession of the various communities.

Pelagic and benthonic organisms

The coastal belt between the levels of high and low tides is called the tidal zone, or the littoral zone. Below it the true submerged world begins. The sea-bed, starting from the low-tide mark, slopes more or less gently until it reaches a depth of from 400 to 1000 feet (120 to 350 metres), usually around 650 feet (200 metres); beyond this the gradient increases sharply. The gentle slope

This view of Bonifacio, Corsica, shows how erosion has affected the cliff, the surface of which has been cut away at a sharp angle; the buildings are perched on an overhang

is known as the continental shelf. The steep slope is the continental slope. The seas that cover the continental shelf are termed epi-continental, and their extent obviously varies with the extent of the continental shelf itself, which is generally greater where mountains are far from the coast, and is at a minimum in places where the land rises sharply from the sea. The whole of the epi-continental waters constitutes the neritic province, and beyond the continental shelf the great mass of water filling the ocean basins constitutes the oceanic province.

Marine organisms are classified as pelagic and benthonic. The pelagic division comprises the whole volume of water, at all depths, that forms the seas and oceans, while the benthonic division occupies the whole of the sea bed, from the shore to the deepest abysses of the sea bottom. The populations of the latter, collectively known as the benthos, include all the sedentary, attached and burrowing animals and plants that dwell on or within the sea bottom, living in constant relationship with it.

The plants and animals of the pelagic division fall into two principal groups, the difference being based on their powers of locomotion. One group, the plankton, merely floats in the water, and thus tends to drift with the tides and cur-

rents. The swimming powers, if any, of planktonic animals and plants serve more as a support, that is, a matter of buoyancy, rather than a means of propulsion. Some planktonic organisms can swim, but not strongly enough to oppose the currents. By contrast, the other group, the nekton, is composed exclusively of animals capable of moving from place to place independently of the flow of water. Nektonic animals include the squid, fish and whales.

As may be expected, a few species of animals and plants do not fall neatly into one of these two categories, the pelagic and the benthonic. Among marine plants, for example, certain species of algae are to be found both on the sea bottom and in the open waters, notably the *Sargassum* weed which is fixed to the sea bed in the Gulf of Mexico, and is therefore benthonic, and is pelagic in the Sargasso Sea, floating with the currents. A similar overlap is evident in the case of animals like molluscs, annelids, coelenterates and certain species of fish. However, these must be regarded as exceptions, because a far greater number of species conforms rigidly to the classification, notably the seaweeds among the benthos, and the comb-jellies (ctenophores) and whales (cetaceans) among the plants and animals of the pelagic division.

The southernmost tip of Africa, the Cape of Good Hope, has for centuries been an important landfall for sea travellers

Where land meets sea

The influence of the expanses of sea is felt over vast areas of dry land, where it conditions the climate, affects the winds and the atmospheric humidity, and determines the nature of much of the vegetation. This influence on land increases progressively as the coast is approached, ultimately creating at the coastal land an environment which owes its distinctive nature to the proximity of the sea. The vegetation of coastal land is predominantly halophilous (salt-loving), but there are also some xerophilous plants (plants able to withstand dry conditions).

Just as some xerophilous plants have made the adjustment to the coastal land environment, so too have the sand plants, the psammophiles (sand-loving plants). One of the important factors that makes this adaptation possible is the saltiness of the marine environment. The remarkable quantity of salt that impregnates the coastal sands has the effect of slowing down the osmotic processes that are responsible for the exchange of water between the environment and the plants, and consequently only those plants that have high protoplasmic concentrations can survive, as they are still able to absorb water through their roots.

Furthermore, the halophilous plants are resistant to dehydration, thanks to a slowing-down of transpiration (loss of water from the surface). Often they have an abundant layer of fine down on the leaves and stems, or a thickening of their encasing tissues which makes the surfaces of these plants impermeable to water. With such a construction, the aerial organs of plants, the leaves and stems, can store considerable reserve quantities of water, as an insurance against prolonged drought. Since these aerial organs become succulent and potentially attractive to plant-eating animals, the stems, leaves and branches of halophilous plants tend also to be tough and fibrous, even spiny.

In addition, these plants need to adapt to the mobility of the fine sediments that make up the soil and to the mechanical action of the wind, which may raise spiral swirls of sand to scour the aerial parts of plants or may uncover their roots. In the wind, everything becomes encrusted with brine and almost buried in the sand that accumulates on the plants. Against the changing moods of the wind, the plants develop long roots, and underground and surface stems, and they reproduce vegetatively as well as sexually.

Such plants provide a considerable defence against the reduction of large coastal tracts to desert, a threat which is partly due to the shifting of the sands under the action of the winds that blow off the sea. Against this, various species of coastal vegetation act not only as windbreaks, but also as stabilizing agents of sand dunes, by binding the sand with their tangle of roots.

Plants are not the only form of life present in the coastal land. Many animals flourish in this environment, and there are occasional visits from species that belong more properly to the hinterland, bringing about an overlap between a completely marine fauna and a wholly terrestrial one. These visitors include various rodents, especially rats which scavenge the shore at low tide, usually at night. Otters readily take up residence on the coast, particularly in temperate regions where winter brings frozen rivers. Others, like the rats, invade the shore for carrion, a notable example being the brown hyena of South Africa, which has been given the alternative name of strandwolf. There are also birds that visit the shore, while normally living inland from the coast: examples include waders, such as the curlew, that probe in mud for a living, and scavengers like certain members of the crow family. Some lizards find the sandy coastlines to their liking, preying upon the many kinds of insects that find a habitation there.

Groups of gastropod molluscs are frequently

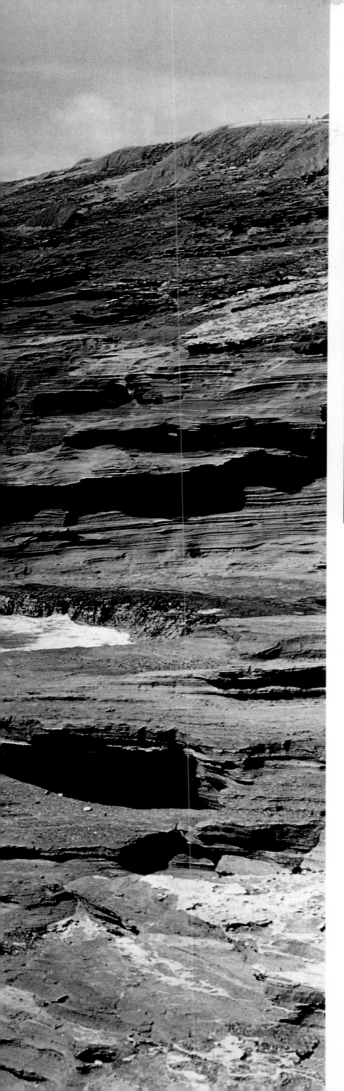

to be observed on the plants of the sandy shores, particularly snails of the genus *Helix*. In the part of this zone nearest the water, one sometimes can find animals that originate from the sea and merely seek out the coast to rest or to breed, the most obvious examples being the seals, and the marine turtles. Other species, like sea-gulls of the genus *Larus* and the terns or sea-swallows, genus *Sterna*, frequent both the expanses of sea and the coasts, as well as lagoons and estuaries. Gulls may move inland during the winter.

The muddy shores support large populations of waders. There are many birds that wade on the shores of lakes or in estuaries at low tide. But the term 'wader' is reserved by zoologists for a large group of birds which vary from small to medium-sized, with generally inconspicuous plumage, long or longish legs relative to body size, and a long or longish bill adapted for probing in mud or soft ground for small invertebrates, such as worms. They include the plovers and many plover-like birds belonging to the order Charadriiformes – more particularly, to the family Charadriidae.

One wader draws attention by its unusual behaviour to a characteristic of the no-man's-land between sea and land. This is the turnstone, so named for its habit of flipping over stones and pebbles with its bill, to eat the small invertebrates sheltering under them. A turnstone will also spend hours on end turning over the semi-dry seaweed that constitutes the line of drift material and flotsam that runs parallel with the water's edge. Although dry above, this line of assorted debris is usually damp underneath and harbours small crustaceans, such as sandhoppers. It can be putrefying, forming suitable material in which flies and other insects lay their eggs. All this also makes a rich feeding ground for small songbirds that invade the coastal edge and the littoral zone.

There is a further smudging of the boundary between the marine and terrestrial environments, if only as a temporary event. For example, on one occasion a European grass snake, *Natrix natrix*, was observed 20 miles out to sea and still heading for the open ocean. There is a surprising number of records in the annals of zoology of other terrestrial or semi-aquatic reptiles, in all parts of the world, seen doing much the same. Whether these have returned to their proper habitat in the end is not known. If they have they are only emulating the behaviour of the estuarine crocodile, *Crocodylus porosus*, of South-East Asia to Australia, including the East Indian archipelago. This, the largest living crocodile (up to 20 feet long), lives

Above: Dry rock pools on a reef above sea level are filled only with brackish or rain water

Left: Cushions of a species of Labiatae *on a rocky spur in the Gulf of Taranto, in south Italy*

Right: The agave, Agave americana, *is a xerophyte that thrives in dry sandy regions because of its ability to store water in its fleshy leaves*

in rivers, estuaries and the sea, and even goes island-hopping, swimming long distances in order to do so.

In reptiles there is no major adjustment needed for a change from fresh to salt water, but marine reptiles generally have specialized glands for the secretion of excess salt. Some fishes also pass from salt to fresh water, and vice versa. Lampreys, sturgeons and mullets do so, while most eels spend the greater part of their lives in fresh water but go to the sea to breed, the reverse of the salmon, which feeds and grows in the sea but enters rivers to breed.

There is less interchange in plants, although lichens can exist in lagoons and briny pools, where a large number of plants live, needing varying amounts of salt to remain there. There are some that flourish in an environment of high salinity.

Obviously, on a rocky coast pools are very frequently formed, some of which will be above the level of the highest tide, others within the tidal zone and thus only uncovered at intervals. These pools are veritable aquaria, in which animal and vegetable life flourishes despite frequent variations in the temperature, salinity, density, oxygen content, and hydrogen ion concentration of the water. The highest pools of the

Right: Vegetation on land behind sand dunes is often immensely varied, due to the fact that both fresh and brackish water is to be found there

The rocky coast of Puglia, near Salento, in Italy. The depressions in the rock may be filled with fresh water – as after rain – or with sea water, from spray. Animals living in the pools must thus be able to cope with a range of salinities

Rocky shores are often richly covered, as this one, with seaweed. Present here are Ralfsia verrucosa *and* Cystoseira stricta

The National Park, Tierra del Fuego: gulls congregate on the shore of the Beagle Channel

reef are topped up by sea water only during periods of heavy seas. Conversely, they are filled with fresh water during rainy weather. Those within the tidal range are alternately submerged and uncovered as the tides dictate. These pools may contain tiny beetles, and adults and larvae of *Culex* and *Aedes* mosquitoes, plus a wide variety of totally marine animals and plants. Often living between one pool and another are lichens and algae, the former being an example of 'terrestrial' vegetation which has adapted to the intermittent immersions, and the latter of marine

vegetation tolerant of frequent exposure. Young crabs dart in agile fashion from pool to pool; and in the tidal pools, even during the ebb, there are various animals that frequent the tidal zone, occasionally leaving their normal habitat to move higher up, but which are equally quick to go down again if faced by adverse conditions.

It is not always easy in practice for the naturalist to distinguish between the two types of pool, the rock pool and the tidal pool, both of which support such a wide variety of life forms. Some organisms live as much in the air as submerged, a good example being the sea snails *Littorina*. The periwinkles *Littorina punctata* and *L. neritoides*, are no bigger than a pea and have a globoid and spiral shell. During winter and spring they cover the rocks, but during summer they retreat into the fissures and into the deepest crevices of the rocks, seeking coolness and moisture. These molluscs can survive in air for many weeks, their anatomy enabling them to adapt in two important ways. The first follows a pattern that is common both among the animals of the exposed coasts and among those of the tidal zone. They have an operculum which seals the opening of the shell and thus enables the animal to retain a supply of water in contact with the respiratory apparatus, or gills, for a long time. This permits the intake of oxygen for a while even when the operculum is closed. The second adaptation consists of a rudimentary respiratory apparatus on the internal face of the walls of the mantle. This system, constituted by a delicate network of blood vessels, allows the absorption of atmospheric oxygen.

This special organ appears to be the precursor of the better-developed respiratory organ of freshwater and land snails. Thus, the periwinkles have a considerable biological interest, as well as being good to eat.

The four species of periwinkle living on the coasts of Europe are equally interesting in the variety they show in their methods of reproduction. Periwinkles, like most marine snails, though with the exception of some limpets and the top-shells, do not shed their genital products, the ova and sperm, into the sea for random fertilization. The male impregnates the female, using an intromittent organ, or penis. Fertilization is, therefore, internal.

In the rough periwinkle, *Littorina saxatilis*, the fertilized ovum completes its development within the body of the female and is born with a shell, resembling the parent except in size. So there is no planktonic larva. The flat periwinkle, *L. littoralis*, lays its eggs in small masses of jelly on the surfaces of seaweeds. In due course they emerge from this, shelled like the young of

Below: Yap Island, a Pacific island situated between the Philippines and the Solomon Islands

L. *saxatilis*. Again, there is no planktonic larva. Largest of the four is the common periwinkle, *L. littorea*. The female of this species liberates a number of horny capsules, each containing three eggs. From the eggs hatch free-swimming larvae that spend a short while in the plankton before settling on the bottom as young periwinkles.

These three species live between tide-marks. The fourth species, the small periwinkle, *L. neritoides*, is characteristic of the splash zone, although it may sometimes be seen lower down the beach. It is, therefore, wetted only at the highest of the high tides and can be said to be almost independent of the sea for day-to-day living, and to be on the way to becoming a land snail. Like all periwinkles, it is herbivorous, but instead of feeding on seaweeds it feeds largely on lichens. This also can justifiably be taken as another step towards becoming a terrestrial species. Yet in its reproduction it is more tied to

the sea than any of the other three species. It spawns during the winter from September to April and it does so every two weeks, coinciding with the spring tides, the only times when the small periwinkle is submerged, especially in winter when storms increase the height of the tides. The larvae are planktonic and spend a long time as such before settling lower down the beach, to change into the adult form. They then creep up to beyond high-tide mark, to the position typically occupied by the species.

Equally interesting are those organisms that populate an environment which, surprisingly, has received relatively little attention; the marine interstitial water in the interstices between the sand grains. This water may be only partly

Tidal rock pools such as this one are very often linked directly with the sea by small channels. In such cases the pools do not contain any special groups of animals

marine, and may occasionally, though rarely, undergo considerable variation in salinity. Besides this, it is often somewhat poor in dissolved oxygen but rich in organic substances.

The interstitial fauna varies in richness according to the locality, and includes vermiform invertebrates of different groups and various crustaceans and insects. It is interesting to note that some are present both in the marine interstitial environment and in the freshwater environment along the banks of some rivers, waterfalls or lakes.

High and low tides

Organisms of very diverse groups flourish in tidal zones. In the Mediterranean, the change in sea level through the tides may be as little as fifty centimetres or even less, while in other places it is much more and in some parts of the world may approach 50 feet (15 metres). In the Bay of Fundy, Nova Scotia, it is almost 60 feet (about 18 metres). There, the sea withdraws from the shore for some miles at low tide, and when it comes in again it creates at times a real sea surge, frequently bringing about a transformation in the face of the countryside itself, submerging much of the coastal land. The effect of the tides is just as noticeable at low ebb: islands are reconnected to the land, while exposure of expanses of the sea bed reveals a luxuriant mantle of vegetation and animal life.

The organisms of this zone pass during the low tide period from an aquatic environment to a terrestrial one, unless they are in a persistent rock pool or stay in the interstitial water of the sandy beach. For the most part they lose water and can thus be partially dehydrated, the more so as in general the salts are concentrated in such water as remains, as a result of evaporation. At other times, by contrast, the sea water of the rock pool is diluted by rain. The air temperature also changes, rising during the day and falling during the night and in spells of bad weather, while other environmental factors may undergo variations that are equally abrupt in character.

The plants and the animals of the tidal zone are so perfectly adapted, frequently through millions of years, to the variable nature of these environmental conditions that to some the conditions have become a necessity of life, not something merely to be tolerated.

Since the higher part of the tidal zone is exposed for longer periods than the lower part, the different species are distributed in a vertical manner through the zone, from extreme low-water to extreme high-water mark, in such a regular succession as to provide well-defined sub-zones, each characterized by the presence of particular species.

The big tides of the English Channel, such as that on the coast of Brittany, uncover distinct sub-zones: beneath a band of lichens flourishes a brown alga, *Pelvetia canaliculata*, which dies if kept in water for long. Below the *Pelvetia* band is another brown alga, and then a band of bladderwrack, *Fucus vesiculosus*, and then one of *Ascophyllum nodosum*, and so on until the lower limit of the tidal zone is reached. There the vegetation is made up of laminarians, or oar-weeds, brown algae which are able to withstand only brief periods of exposure.

Along the coasts of the Mediterranean, the extent of the tidal zone is limited by the limited size of the tides. Two horizontal levels can be distinguished, however: the upper one, characterized by longer and more frequent exposure, is populated by the algae *Bangia fuscopurpurea*, *Porphyra leucosticta*, *Rissoella verrucosa*, and *Lithophyllum tortuosum*, while the lower one, receiving rarer and shorter exposures, is populated by the algae *Nemoderma tingitanum* and *Corallina mediterranea*. Below the tidal zone extends the submerged reef, for the most part very rich in algae, but these, such as the brown alga *Cystoseira mediterranea*, will die if subjected to relatively short periods of exposure.

The two horizontal levels of the Mediterranean tidal zone are not, however, equally apparent on all the coasts, and in certain localities algae appear in locations which differ from those cited. *Fucus virsoides*, for example, grows in the upper Adriatic, where the size of the tide is considerable for the Mediterranean regions.

Many animals live among the algae on the reefs where the water is relatively pure. Particularly evident are highly modified crustaceans, the acorn barnacles *Balanus* and *Chthamalus*, limpets (*Patella*), some sea urchins, like *Arbacia lixula*,

Rock pools on a rocky coast can vary in salinity daily, according to whether it is raining or the sun is shining while the tide is out

Left: The many-coloured fronds of the seaweed Cystoseira stricta, *or rainbow bladder-weed. Below:* Balanus perforatus, *one of the acorn barnacles, growing on rocks in the littoral*

and coelenterates, like the sea anemone *Actinia equina.*

The acorn barnacles are easily recognized since they are protected by a shell in the form of a truncated cone adherent to the rock at the base and with cirri emerging from the top when the tide is in. During exposed periods, these animals close the aperture of the shell with a movable operculum composed of four plates, trapping within the shell an amount of water which may be enough for respiration until more is brought by the next high tide. If not the operculum may open from time to time and limited aerial respiration occur.

The acorn barnacles *Chtamalus stellatus* and *Balanus tintinnabulus,* which resist the battering of the most violent seas, are common on the reefs around European coasts.

The limpets of the genus *Patella* have a shell in the form of a very flattened cone, frequently covered with very small algae. They are difficult to detach from the rock unless they are taken by surprise. When menaced they adhere so tenaciously that they become almost a part of the rock.

The red sea-anemones that during the period of low tide look like rounded bags of jelly, appear as cylindrical sacs crowned with tentacles and slowly take on the appearance of chrysanthemum blooms when covered by calm water. If a wave

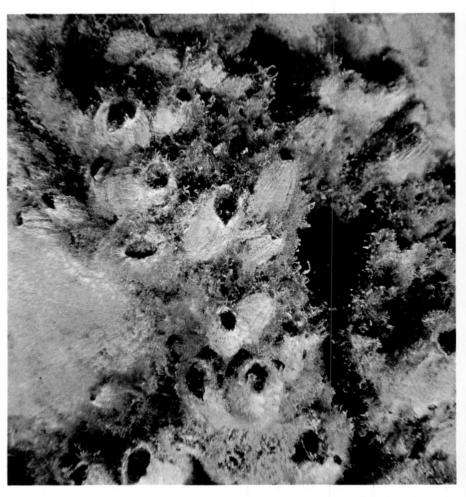

Low tide at the island of Mont St Michel, off the coast of Brittany, is made more impressive by the dramatic recession of the sea over the sands

Right: Low tide exposes a rocky reef covered in seaweed, running along a beach that is almost totally devoid of vegetation. Below right: A luxuriant growth of seaweed exposed at low tide on the coast of Liguria, Italy

Left: The littoral zone at low tide; the vegetation is typical of many European Atlantic shores

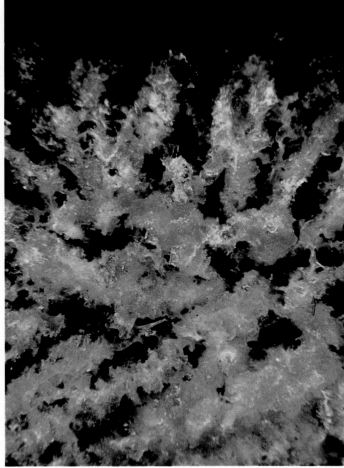

laps them gently, these sea anemones undulate with the water while the tentacles move around in search of food, but when thrashed violently they contract, taking on the appearance of a tomato. Limpets and sea anemones are able to creep on the rock, but only very slowly.

It usually comes as a surprise to learn that sea anemones are mobile. Perhaps their name, linking them with plants, automatically characterizes them as completely sedentary. Although some species stay indefinitely in one spot provided circumstances are favourable, most move about readily should occasion demand, and some are seldom still. The method of locomotion is varied. The most common is brought about by move-

ments of the basal disc similar to those seen in the foot of a snail. A few species drag themselves along by means of ridges running along the sides of the body, slug-like, with the crown of tentacles in the van and the basal disc to the rear. Some let go their hold on the substratum altogether, inflate the body and are carried about by the currents, becoming temporarily pelagic. Others, using the same method, float head downwards on the surface. Some sea anemones can walk upside-down on their tentacles. In very rare instances, sea anemones have been seen to progress by movements strongly reminiscent of jumping. In a few species, lashing movements of the tentacles approximate to swimming.

Submerged reefs

The submerged reef, below low-tide mark, is frequently clothed in algae, which tend to become progressively more luxuriant as the water becomes clearer, purer and more sheltered from the violence of heavy seas. Just as progressive changes are evident in the land flora with increasing altitude, so marine vegetation changes as the reef goes deeper into the sea. The major difference is that on land a significant variation of the flora requires a change in level of hundreds of feet, whereas a comparable variation in the sea requires only a few feet: in fact, the surrounding conditions in the sea vary more strikingly with the depth than those on dry land with the altitude.

In some places along a submerged reef one alga predominates over all the other species, and this confers a known characteristic upon the landscape, just as happens on dry land, where in a field or in a forest one species of vegetation prevails in a significant manner over all the others. But often in the meadows, pastures and scrublands many forms of plant intermingle without one being conspicuously more abundant than the others, and each season brings something into bloom, while others are withering. In the course

Acetabularia acetabulum, *one of a genus of green algae remarkable for their resemblance to small toadstools.*

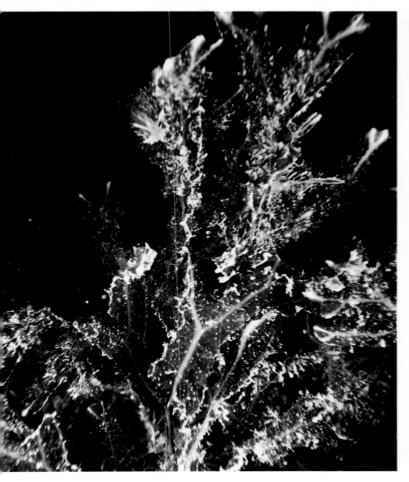 

of the summer the same field or meadow undergoes continuing changes in colour. Then in
winter, it appears that plant growth is arrested,
with the colours becoming uniform and dull.

The majority of the marine plants are algae
and thus show no flowers, but even these exhibit
in the course of the year a period of full development and one of regression. At the end of autumn
the rocks that were earlier a mass of luxuriant

vegetation are more or less bare; few algae can
be seen, and then only of a small size. But towards
the end of the winter, as if in anticipation of what
will happen to the vegetation on the exposed
land, the vegetable life in the sea along the submerged reef begins to extend, starting slowly at
first, in January, and then expanding progressively and more luxuriantly.

On some reefs in autumn a reddish colour predominates, due to the presence of red algae. That
basic colour remains dominant during a large
part of the winter, but towards spring the reef
begins to be masked by brown algae.

In full summer the green alga known as sea
lettuce, *Ulva lactuca*, covers shallow bottoms of
the sea almost like a vegetable garden, and is frequently mixed with another green alga, *Entero-
morpha compressa*, with its long green ribbons.
A third green alga, *Cladophora lanosa*, forms a
slippery skin over the rocks.

During the summer season, the whole reef is
covered with algae of many kinds of ever-
changing colour and form, appearing like a living
colour palette. Many institutes of marine or
botanical research possess collections of marine
algae, and there is no doubt that these plants are
among the most beautiful and interesting of all,
apart from their biological importance.

The vegetable life often appears to be dominant on the submerged reefs, but this vegetation is pervaded by a host of animal life, since each alga harbours its own animal or animals. One may carry a small colony of hydroids that covers the stalk of the alga like a moss. But if the kinds of algae are numerous, so too are those of the hydroids and the other inhabitants that live on the algae – like the bryozoans, the worms, crustaceans, and molluscs.

One well-known worm is the sedentary *Spirorbis*, its tiny tube coiled like a mollusc shell. It is an aberrant member of the marine bristle-worms. Creeping among the algae are various sea slugs, animals as exquisitely beautiful as land slugs are dull. At the bases of the seaweeds move various crabs, including the spider crab that camouflages itself by plucking tiny pieces of algae and accommodating them on minute hooks on its own back.

A wide variety of animals is found where stony shallows are formed. These include the breadcrumb sponge (*Halichondria panicea*), spider crabs with legs so long they give the appearance of a spider's web, the starfishes and brittlestars and tube-building marine worms.

Left: A further view of the sea anemone Actinia equina, *this time with its tentacles expanded*

Left: A Mediterranean sea anemone with clubbed tentacles

Above: Starfishes or sea stars. Hacelia attenuata *on the left,* and Ophidiaster ophidianus *on the right*

Right: The Mediterranean sponge Petrosia ficiformis, *with the starfish* Coscinasterias tenuispina

The waters around the submerged reef contain myriad forms of animal life, many being characteristic of certain depths while others are capable of living in or around the reef, without being dependent on conditions at any given depth. Again, some forms encountered in the submerged reef are also familiar in other zones, moving freely between them as the seasons change.

The various species found on the surface include the small cushion star (*Asterina gibbosa*), various sea cucumbers, tunicates or sea squirts such as the solitary *Ascidia mentula* and *Ciona intestinalis*, colonial tunicates such as *Botryllus schlosseri*, and small vase-shaped sponges like *Sycon ciliatum* which, in one form or another, seems to abound in shallow seas almost all over the world.

From about three feet below low-water level the characteristic forms are echinoderms, for instance the sea urchins which live at depths of less than 30 feet (10 metres).

Some tunicates live at greater depths, like the red or rosy *Halocynthia papillosa*, and the sea lemon, *Microcosmus sulcatus*, as well as madrepores such as *Astroides calycularis* and sponges like *Axinella polypoides*. Some animals already mentioned in connection with the tidal zone are

Above: Codium vermilara, *a small green seaweed that grows in tufts. Left:* Codium bursa, *common in the Mediterranean, growing here with hydroids on* Eunicella

found on the upper part of the reef, examples being the limpets, the acorn barnacles and the 'sea tomato', the contracted red sea anemones. This part of the reef is also the home of the snake's locks sea anemone, *Anemonia sulcata*, and a wide variety of sea snails, notably in subtropical and tropical seas, the cone shells and the conches. During the summer in subtropical seas, some of these sea snails move into the tidal zone and may survive in the air for fairly long periods.

Other gastropod molluscs abound in this zone, the most interesting being the sea hares, *Aplysia*. These are hermaphrodite molluscs which possess only the vestige of a shell. Their eggs are sometimes called 'sea tangles', because of the entangled appearance of the cords attached to the egg-capsules which cling to the pebbles and to the submerged rocks as the eggs are deposited.

Bivalve molluscs represented on the submerged reef may include the oyster (*Ostrea edulis*) with its delicious flesh, the spiny oyster (*Spondylus gaedoropus*), the saddle oyster (*Anomia ephippium*) which adheres firmly to rocks or to shells with one of its valves, and the variegated scallop (*Pecten varius*).

Mussels, much appreciated as 'sea food' in some parts, are widespread over a submerged reef. They nestle in the rocks, hollowing out

galleries for themselves by means of acid secretions, a habit indulged in especially by the mussels known as 'sea dates' (*Lithophaga lithophaga*). Other bivalve molluscs are the piddocks (*Pholas dactylus*) which burrow into the rocks with a rocking movement of their valves, and which are luminescent because of a secretion of luminous slime from glands situated on the siphons and on the sheath. Viewed in darkness

the outline of these creatures is marked out with a greenish light.

From the surface down to as much as 800 feet (about 250 metres) in depth, the common octopus (*Octopus vulgaris*) is to be found. This well-known cephalopod mollusc may sometimes leave the water for a short time, if anything which is outside of the water has attracted its attention. Under pebbles and stones, even at the most shallow depths, can be found the chitons, (e.g. *Acanthochitona*), which are oval molluscs with eight shell-plates, and a gastropod, known as the ormer (*Haliotis*), whose shell is shaped like the human ear and is studded along the edges by holes. This is a refinement which allows water to leave the mantle cavity.

On pebbles or old shells on the reefs are found tube-dwelling marine worms like the serpula, *Serpula vermicularis*, and peacock worm, *Sabella* (*Spirographis*) *spallanzanii*, a species which also lives in the stony shallows and on various types of sea bed. Single or small bundles of twisted whitish tubes often appear on the reef, reinforcing those of the marine worms. These are the shells of aberrant gastropod molluscs of the genus *Vermetus*.

Crustaceans also abound on the reef at a certain depth, particularly the crawfish or spiny lobster *Palinurus vulgaris*, the lobster *Homarus vulgaris*, the sponge-carrying crab *Dromia vulgaris* (which frequently carries a sponge on its back, such as the sea-fig *Suberites ficus*), the hairy crab *Dorippe lanata*, the crabs *Scyllarus arctos* and *S. latus*, and a whole series of little crabs, the most common of which is the prawn *Palaemon* (*Leander*) *serratus*.

The whole submerged reef provides a home for a great number of fishes, the same species being found where the vegetation is dense as where it is sparse. Although many do not readily leave the reef, some can be caught on other types of sea bed. In addition, similar forms are common in the beds of kelp or on sandy bottoms.

A common sight on the reefs are the blennies and gobies, the latter adapted to life among stones and on the rocky reefs. This adaptation takes the form of fusion of the two ventral fins to form a sucker for adhering to solid surfaces. A similar adaptation is exhibited by another inhabitant of stony bottoms, the shore clingfish or Cornish sucker (*Lepadogaster lepadogaster*).

In shallow waters, the bass known as the comber (*Serranus cabrilla*) and *S. scriba* are common, while at greater depths various species of stone bass or wreckfish *Polyprion* and sea

Above: A cushion star, Asterina gibbosa, *common in European waters*

Left: The sea hare, Aplysia, *is so called because of its body lobes, which resemble the ears of a hare*

perches *Epinephelus* are prevalent. The wrasses, which prefer the marine meadows, but often also move on to the reef, are represented by forms of the genus *Crenilabrus*, such as *C. mediterraneus*, the rainbow wrasse, *Coris julis*, and some species of the cuckoo wrasse genus *Labrus*.

Many sparids move on to the reef, among them the bogue (*Boops boops*), and it is not rare to find mullet also visiting this zone.

Not all these fish can be described as permanent members of the reef community, since many of these highly mobile animals wander from one bottom to another, 'passing through' the waters of the reef as they journey to the neighbouring expanses of sand and seaweed beds.

It is appropriate at this point to discuss the subject of *luminescence*, or living light, as it is sometimes called, especially as repeated references are made to it in later pages. The phenomenon is widespread among animals and may take a variety of forms. One of its most significant features is that it is a cold light, meaning that there is very little loss of energy as heat. If it were possible to probe the secret, it would be possible to produce an artificial illumination for our own purposes which would be more highly efficient than our present form of lighting

Above: The common Atlantic octopus

Left: Blennies of the type illustrated here are very common on submarine reefs, and are also to be found in rock pools when the tide is out

Far left: The seaweed Padina pavonica, *commonly known as the peacock's tail seaweed. Near left:* Halimeda tuna, *a green alga whose tissues have become impregnated with lime salts. Below left:* Serranus scriba, *one of the sea basses*

Top right: Palaemon (Leander) serratus, *the common prawn. Below right: The cuckoo wrasse,* Labrus bimaculatus

A further view of the
cuckoo wrasse, this time
the female, who differs
remarkably in colour
from the male

and less expensive. A good deal is known about
the biochemistry of luminescence, but all that we
need say here is that it is due to a substance called
luciferin when it is being acted upon by an
enzyme called luciferase.

Among terrestrial animals luminescence is
best known in the flashing light organs of fireflies
and in the glow-worm. In these, the lights are
used as a sex-attractant. In the glow-worm, for
example, only the female shows a light. She is
wingless and the male would have difficulty in
finding her without the kind of beacon, or some
equivalent signal, which she carries. Among fireflies the lights have specific patterns of flashes
by which the insects can recognize members of
their own species, basically for purposes of
mating.

Luminescence, or bioluminescence as it is now
more commonly called, is more frequent among
marine animals, notably among deep-sea fishes,
many of which are liberally equipped with rows
of photophores, or light-organs, along the sides
of the body, recalling the lighted portholes of a
ship at night. It seems beyond doubt that the
patterns of the photophores serve as species
recognition, both for mating purposes and for
recognizing members of the same species. In
carnivorous deep-sea fishes there is another use

for bioluminescence, as a lure for attracting prey
to the vicinity of the mouth, where it can be
snapped up. There may be the general use,
although how far this is effective can only be a
matter for speculation, in illuminating the stygian
darkness of the ocean deeps.

Clearly the photophores of fishes, and also of
deep-sea squid, are functional. They have a
definite use. It is less easy to say the same for
many of the marine invertebrates, and especially
of those that may light up the surface of the sea
at night when it is disturbed.

To use light, eyes are necessary, and many
species that are luminescent have no eyes. It may
be their luminescence is defensive, but it is more
easy to postulate this than to support it by
reasoned argument. It is significant that so often
these animals light up when disturbed, as when
the surrounding water is agitated. It sometimes
happens that in walking along a sandy shore at
night, when the tide is out, one's footsteps leave
a glowing print which slowly fades, caused by
micro-organisms which have perhaps been
stranded by the ebbing tide. The greatest mystery
of all in the use of bioluminescence must surely
be that a mollusc, like the piddock, should bore
into a rock for shelter and then betray its
presence by glowing at night.

Coralligenous formations

A coralligenous bottom or formation is a descriptive name for submerged reefs lying offshore and built by other organisms than the true or reef-building corals. They might equally well be called false-coral reefs.

In the books in English dealing with the seabed or with marine zoology generally, they are usually given little attention compared with the true coral reefs. They tend to be included under the heading 'Coral and other reefs' and the pages that then follow are devoted mainly to coral reefs. In ordinary terms, also, their identity is masked because they are often called, erroneously, coral reefs.

Large reefs, particularly in the West Indies, are known as coral reefs although they are built up largely by hydrocorallines of the genus *Millepora*, the fire-coral, a close relative of the sea-firs.

Tube-building bristle worms and others secreting a limy tube or skeleton also contribute. Even algae take part in their formation, the most common being the lime-encrusted *Lithothamnion* that may form miniature 'corals' of a few centimetres height.

The surfaces of the submerged reef, at depths of 65 to 650 feet (between 20 and 200 metres), are covered with a mantle of plants and animals of the most varied groups. From a depth of about 65 feet (20 metres), the waters that cover these plants and animals are fairly calm, cool and semi-dark; where the water is constantly clear and free of any pollution it allows the development of characteristic coralligenous communities.

The name used here, coralligenous formations, indicates the presence of corals, although the red coral, *Corallium rubrum*, now appears in relatively few localities, while the formation with which we are concerned is widespread. *Corallium rubrum* produces the red precious coral which, in the past, has been in much demand for jewelry.

Axinella cannabina, a branching sponge with a siliceous skeleton

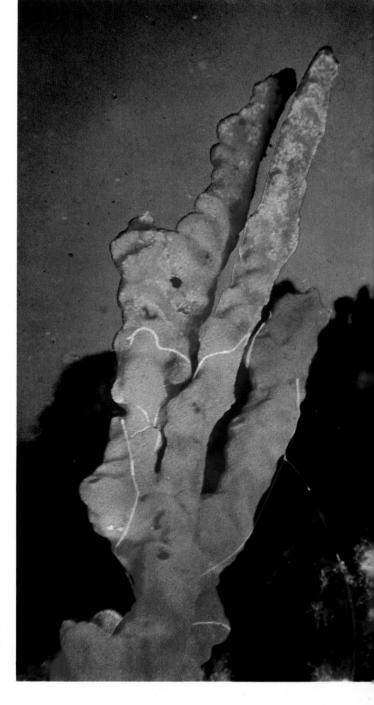

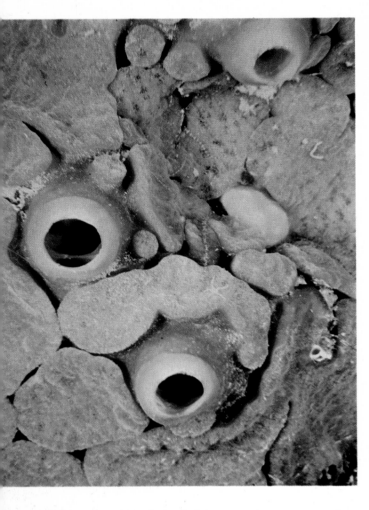

often referred to as being 'pre-coralligenous'.

On the other hand, where the conditions of the water allow the establishment of these formations, any stable substratum with many rough edges may be studded with a number of coralligenous elements, providing that it is of certain minimum dimensions and at a variable depth. This is true both of substrata that project from an expanse of sand or mud and of those that are

Below: A tubular, non-commercial sponge with a fibrous skeleton; Callyspongia vaginalis

Divers have exploited this to such an extent that the coral has been wiped out in places where it was once abundant. Another denomination of the same nature is that of coralline bottoms, due to their constant richness in corallines, calcareous red algae, not all of which are, however, exclusive to this type of community.

The most important environmental factors for the development of the coralligenous biocenoses seem to be the presence of a stable substratum, of no matter what kind, and variation within narrow limits of solar illumination, water movements and temperature, chemico-physical composition and purity of the waters, rather than considerations of depth and factors correlated with it. It follows that a coralligenous formation can develop even at the most shallow depths – in partially submerged grottoes, on platforms and on rocky overhangs, and even in the remains of sunken ships. Further, while some animals and plants are exclusive to the coralligenous formations, others are not, even if they are prevalent within them. There are some anomalous formations that are not truly coralline (since they lack some of the important biological elements that are characteristic of the coralligenous formation), but are to a lesser or greater extent akin to coralline; these formations are therefore

*A branch of red coral
with its polyps extended*

to be found isolated between beds of seaweed.

The substratum can be composed of rocks and of solid bodies of a varied nature, but not, as a general rule, of unimpacted sediment, such as sand or mud, since the species of plants and animals that constitute the fundamental part of the coralligenous formation cannot anchor themselves in loose material. Among the plants that make up this formation are those of the genera *Lithophyllum, Melobesia, Peyssonelia,* and *Pseudo-lithophyllum (Crodelia),* all calcareous red algae

that form layers and encrustations on the substratum, which is further accreted with sponges, madrepores, hydroids, bryozoans, tube-forming bristle worms and molluscs. Its appearance is of a very spongy calcareous mass, full of sharp edges and cavities, in which is a whole world of minute animals able to insinuate themselves into the smallest of spaces. It is on these colonies that sea urchins, starfishes, ragworms, gastropod molluscs and fishes move around. The coloration of the coralligenous formation is basically

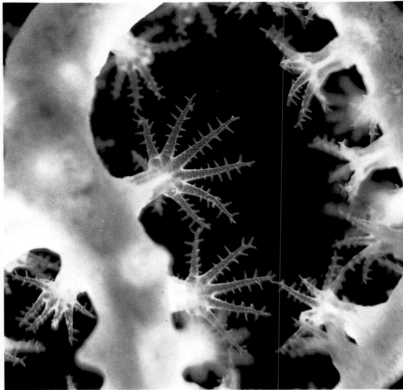

Top left: Foliaceous red algae of the genera Peyssonelia *and* Kallymenia *encrusting undersea rocks. Top right: Polyps of* Eunicella stricta, *one of the soft corals. Left: A slime sponge,* Oscarella lobularis. *A bove: Polyps of the soft coral,* Alcyonium coralloides. *Facing page: Polyps of* Astroides calycularis, *a stony coral that lives just below low-tide mark*

whitish, roseate or violet, but each of its parts is a microcosm of organisms of the most varied forms, exhibiting a vivid display of contrasting colours. Unfortunately, many of these colours cannot be distinguished without artificial light, since the solar illumination is inadequate at this level.

Certain animals are characteristic of the coralligenous formations. Among the sponges are the sea kidneys (*Chondrosia reniformis*), and the beautiful species of *Axinella* and also *Petrosia ficiformis*. Among the characteristic anthozoan coelenterates of the coralligenous bottom are the unstalked alcyonarian (*Alcyonium acaule*) related to the 'dead man's fingers' (*A. digitatum*), and *Alcyonium coralloides*, all of which form red structures. The madreporites are numerous and include the sulphur-yellow *Leptopsammia pruvoti* and *Caryophyllia smithi*.

These various species are solitary polyps, each having its distinct and robust calcareous support, even if at times they appear to form colonies. It is, however, a colonial madreporite, *Cladocora cespitosa*, that forms the tufts, often of

A feather star, Antedon, *related to sea lilies of the deep seas but itself an inhabitant of shallow seas*

Right: The egg-case of a dogfish attached by tendrils from each of its corners to a sea-fan

large dimensions, of calcareous cylinders supporting as many polyps of a violet-brown colour. This species is not characteristic of any particular environment, but is abundant on coralligenous formations, down to a depth of about 2000 feet (600 metres).

Present on coralline bottoms, but not exclusive to them, are the orange or yellow *Parazoanthus axinellae*, polyps sometimes found growing on sponges of the genus *Axinella* but also found on rock or calcareous algae, even on lobsters, as well as *Eunicella cavolinii*, a yellow orange 'sea fan'. Both species are also found on reefs.

Related to the *Eunicella cavolinii*, and frequently confused with it, is *E. stricta*, which represents an element of pre-coralligenous communities rather than one of true coralligenous origin. At a depth of a hundred feet or so, the *Eunicella cavolinii* is succeeded by the large *Para-*

Right: Alcyonium digitatum, *commonly known as dead man's fingers, extends its polyps*

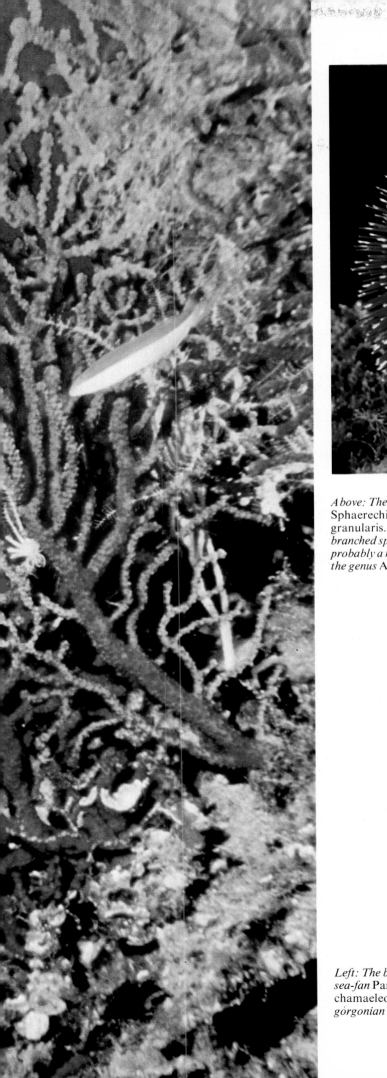

Above: The sea urchin Sphaerechinus granularis. *Right: A branched sponge, probably a member of the genus* Axinella

Left: The beautiful sea-fan Paramuricea chamaeleon, *one of the gorgonian soft corals*

muricea clavata (chamaeleon), which forms veritable submerged 'forests'. These extend to considerable depths, and are found on a variety of bottoms, not just on the coralligenous floor.

Various hydroids are present on the coralline bottom, but are also seen on other bottoms. It has been said that this formation would also merit the name of bryozoan foundation, given the great richness of individuals and of species with which this zoological group is represented on them. The false coral, *Myriapora truncata*, which takes the form of shafts a few inches high is also abundant. It is notable for its dichotomous branches and its roseate colour, which is reminiscent of red coral. Another common species is the sea lace, *Retepora cellulosa*, an elegant species with what looks like a starched rosy lacework.

Among other common bryozoans are the tufty thickets of the *Hippodiplosia fascialis*, which exhibit a characteristic foliate form and a colour that resembles that of the false coral, and the yellowish horned branches of *Porella cervicornis*.

The sedentary bristle worms make a big contribution to the cementing and consolidating of the base, extending from piece to piece, insinuating themselves between one animal and another and between plant and plant, welding their own tube to the substratum, and wrapping it around

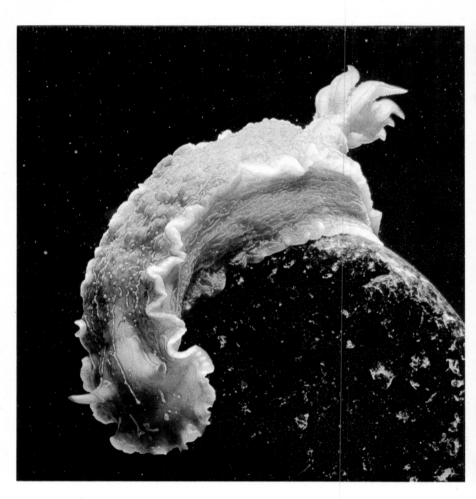

Above: A sea slug, Glossodoris *sp., with its tuft of gills conspicuous at the rear end of the body*

Left: One of the more beautiful of the sea slugs, Peltodoris atromaculata

56

so that it does not move. Within the tube lives the worm which pushes out through the opening as a tuft of tentacles of various colours, sometimes striped, but always bright and delicate in shade. One of the most beautiful is *Protula protula*, with red tentacles that protrude from the head of the twisted white tube. The free-living annelids, however, slide over the tortuous surfaces into the cavities of the calcareous crust.

Other groups of vermiform invertebrates are present on the coralligenous formations: representatives include the nemertines of the genus *Cephalothrix*, and the echiuroids, with *Bonellia viridis* a species of considerable interest. The ovoid body of the female of this species carries an extremely long extensible proboscis, that ends in the form of a T. Along this proboscis is a groove along which food is conducted to the mouth by cilia. The extremely small males are located in a cavity within the female.

Echinoderms exist in profusion in the coralligenous zone: among the sea urchins or echinoids are *Sphaerechinus granularis* and *Echinus acutus*, which are distributed between 65 and 42,000 feet (20 and 13,000 metres), and *Stylocidaris affinis*, distributed in the narrower depth range of 100 to 3200 feet (30 to 1000 metres). These sea urchins belong to the group of

Above: Egg capsules of Thais (Purpurea) haemastoma *clustered on a rock surface*

Left: A tube-building marine bristle worm, family Serpulidae, with its crown of tentacles

the regular echinoids with a subspherical form, straight ambulacra, and anal and oral apertures at opposite poles of the body. Co-existing with these sea urchins on this type of bottom are representatives of the irregular echinoids, species possessing a flattened or discoid body and a dorsal surface that bears the five ambulacral zones, joined together to form an outline resembling that of a five-petalled flower. The anus is located in one of the inter-ambulacral zones, and in some species is so displaced in relation to the dorsal pole that it can be found on the ventral surface of the creature. The oral aperture can also be displaced forward, as in the family Spatangidae. In these echinoderms, the radial symmetry characteristic of the phylum is no longer recognizable, being replaced by a bilateral symmetry. Another irregular echinoid found on the coralligenous formation is the violet-coloured heart urchin *Spatangus inermis*, whose dermal

skeleton is uniformly covered with short spines and with longer spines collected into two groups. Brittlestars or ophiuroids nestle within the mass of the concretions, the most delicate of these being the *Ophiopsila aranea*, while on the surface live the holothuroids or sea cucumbers, and crinoids or feather stars, species of *Antedon* related to the better-known sea lilies which are also found on other types of bottom.

The molluscs of the coralligenous bottoms are said to be legion, and to play a most important part in determining the pattern of the fauna. Conspicuous by their dimensions among the lamellibranchs are the scallops, including the 'pilgrim's shell' (*Pecten jacobeus*), but more common than this are *Pecten flexuosus*, *P. varius*, and *P. opercularis*, some of which are also found on the reefs. The wing shell (*Pteria hirundo*) is characteristic of this zone, attaching itself to the branches of the sea fans, while two other charac-

Left: The delicate seahorse Hippocampus

Above, left and right: Leaves of the marine flowering plant Posidonia *which is common both on beds of sand and deeper in the seas, where it forms a carpet on what are known as the undersea meadows. The fauna that lives on* Posidonia *is complex, its life linked intimately with that of the plant*

teristic species – the rock borer (*Hiatella rugosa*) and the flask shell (*Gastrochaena dubia*) – live fixed in the vegetable concretions or in the rock. Among the gastropod community, the most striking species is the triton or bugle horn (*Charonia gyrinoides*); up to a foot (30 cm) in length, it is the largest European gastropod. Others present are the needle shells (*Cerithium*), the pelican's foot (*Aporrhais pes-pelicani*), the spindle shell (*Fusinus rostratus*), and species of *Trunculariopis* and *Murex*. Again, some of these forms are not exclusive to this type of bottom.

Crustaceans are numerous on the coralligenous bottom; in particular there are the decapods and those crustaceans with shell and limbs sculptured to resemble the bottom. Some species, for example *Pisa corallina*, place pieces of sponge on the back as camouflage. The crawfish (*Lambrus angulifrons*) and the spiny spider crab (*Maia squinado*), the largest of the Mediterranean crustaceans and a popular sea food, are frequent.

With good reason one zoologist has written that the 'fauna of this bottom is rich, but not all the species that compose it are exclusive. In fact the coralligenous formation of the open sea benefits from contact with the coastal mud and with the submarine meadows, whence not a few animals penetrate; the coralligenous formation of the reef, however, draws a part of its population from the surrounding rocky zones'.

This is particularly true of fishes, which can so easily and quickly move from place to place. However, some species are clearly characteristic of the coralligenous formations, such as the lesser spotted dogfish (*Scyliorhinus caniculus*) and the nursehound or greater spotted dogfish (*Scyliorhinus stellaris*), species that resemble sharks in form but are much smaller, not often exceeding a foot (30 cm) in length, and quite harmless to man.

Among the bony fishes in this zone are many species that also inhabit the reef, examples being the moray eel (*Muraena helena*), various bass, and the cuckoo wrasse (*Labrus bimaculatus*).

Among the branching sponges, such as *Axinella* and sea fans, are frequently found seahorses such as *Hippocampus brevirostris*, which temporarily anchor themselves with their tails to the branches or to seaweeds. Nevertheless, the true habitat of seahorses is the submarine meadows and not the coralligenous zone.

Beds of sand

The benthonic fauna as a whole is more closely linked with the nature of the bottom and its plant life than with depth and other environmental factors. Individual benthonic species have a considerable vertical distribution.

Where the coast drops gently away into the sea, the sea bed descends slowly towards the open ocean and is usually sandy near the shore. At greater depths beyond this it becomes first a mixture of sand and mud, and finally pure mud. On coasts of this type the shore itself is made up of long stretches of sand, and the coastal zone, both above high-tide mark and below, is populated by animals and plants and has hydrological changes which differ considerably from those of zones with rocky reefs and cliffs.

The sands mainly originate from inorganic detritus brought down by rivers and progressively spread over long distances by currents, together with particles eroded by the sea from rocky coasts.

Waves fling animals against the rocks, grinding their shells or soft parts and fragmenting and pulverizing them, so that in the sands are varying quantities of shell fragments from cockle and mussel shells and other bivalve shells, pieces of shell of sea snails, and skeletons or fragments of skeletons of sea urchins, corals, bryozoans and other invertebrates. In tropical zones the sands are sometimes so rich in such fragments, particularly reef-building coral, that the shore is white and shining.

At shallow depths, down to 160 to 260 feet (50 to 80 metres), the coastal sands frequently provide a home for a rich benthonic vegetation, made up of seaweeds and some of the higher plants collectively known as phanerogams or flowering plants. Through the whole world there are only about thirty species of these, principally belonging to the genera *Zostera*, *Cymodocea*, *Posidonia*, *Halophila*, *Thalassia*, *Halodule*, *Phyllospadix*, and *Diplanthera*.

Although these marine flowering plants are in some places isolated from each other, they tend to be grouped, forming vast belts known as the submarine meadows. These are particularly evident in the warmer seas, especially in the tropics. Beds of marine flowering plants also flourish, however, off the coasts of the temperate and cold seas. For example, off the Dutch coast lie some thirty-seven thousand acres of sea bed covered with beds of *Zostera*, the eel grass, and the submarine meadows off the Danish coast produce about 24 million tons of fresh plant material per annum – more than the annual production of forage in Denmark itself.

This last piece of information came from a survey made in 1910 by the Danish marine biologist C. G. J. Petersen. In 1931 an epidemic disease struck the eel grass along the Atlantic coast of North America. This soon spread to European waters. The first signs of disease were black spots and stripes on the eel grass, and after this leaves and roots soon rotted. At first it was supposed that this was a bacterial disease. Then a marine fungus, *Labyrinthula*, was suspected, but even now the cause of the dying off of the eel grass is still a matter for speculation. One aspect of the epidemic is that it also reached the Pacific coast of North America, believed to have been carried on oysters from the Atlantic coast put down in Californian waters.

The loss of the eel grass affected a number of animals seriously, for the grass had been a shelter for many kinds of fishes. Many small invertebrates fed on it, including a gastropod mollusc *Rissoa* which, according to the Danish biologist Gunnar Thorson, could be found in hundreds of thousands in each cubic metre of the grass. With the loss of the eel grass some of these species became rare. There was, however, an unexpected sideline.

There were marine biologists who, prior to 1931, had taken the view that organic particles from the death and decay of eel grass constituted a major source of food for many shallow water animals. Surprisingly, although some animals such as *Rissoa*, that fed on the living plant, and the fishes that found shelter and sustenance in the meadows of eel grass, suffered, the majority of shallow water animals living in the vicinity of the meadows showed no adverse effects.

For several years following the loss of the eel grass those who favoured the view that its decay was a source of particulate food held their ground. They argued that the large mass of decaying grass was still fulfilling this function. The tissues of the marine flowering plants are remarkably resistant to ordinary forces of disintegration. Today the results of this natural experiment leave us in no doubt that they were wrong. In many parts of the world the eel grass has never completely recovered. In the Kattegat there are still large areas of sandy bottom bare of eel grass where it was formerly abundant. Continuing surveys of the marine fauna there suggest that, in general, the loss of eel grass has done little to upset the balance of the species.

The submarine meadows in the tropical Atlantic are made up of dense beds of the turtle grass, *Thalassia testudinum*. It is less grass-like, with shorter and broader leaves. The richness of these beds is graphically expressed in figures quoted by Thorson: 113 species of seaweeds using turtle grass as a habitat and 30,000 individual animals per square metre of the meadows.

In the Mediterranean, *Posidonia oceanica* forms a virtually continuous belt of meadow around the coasts, frequently with *Zostera* and *Cymodocea*. Where such meadows are located on a sea bed that is horizontal or slopes only slightly towards the open ocean, they assist coastal sedimentation by detaining and precipitating a large part of the inorganic detritus brought down by rivers. Furthermore, their rhizomes tend to grow upwards as well as horizontally, so that a meadow is constantly rising, on average by about a yard every hundred years.

When the temperature of the surface waters of the Mediterranean is favourable, *Posidonia* reaches up to the surface and multiplies more by vegetative reproduction, through the abundant proliferation of the rhizomes, than through flowers and seeds.

It is important at this stage to distinguish between submarine meadows and the 'forests' formed immediately below the tidal zone by various species of brown algae. There are three major differences between these two types of area: they are established in different ways, they

Two cuttlefishes, the upper one showing the characteristic zebra pattern that is the courting dress of the male

are composed of essentially different floras (the algae being cryptogams), and they shelter entirely different forms of animal life.

Although the beds of *Posidonia* frequently take root on sand, they do not always occupy its entire surface. Frequently, bare sands extend between one meadow and another, giving an irregular patchwork of *Posidonia* and bare sand.

The various types of marine bottom do not always present a regular succession, a fact borne out by certain *Posidonia* meadows. For example, some *Posidonia* beds harbour on their rhizomes a pre-coralligenous biocenosis (characteristic community of species), and extend as far as the base of a reef where they become mixed in with the seaweeds, while others may flourish on sediments not solely made up of sand. This is but one example that upsets any arbitrary classification, although attempts at such classification are not valueless.

Before turning to a discussion of the meadows of *Posidonia*, some attention must be given to the flora and fauna of sandy bottoms. On superficial examination, the sands uncovered by the tide appear to be almost devoid of life, since the animals that populate them either possess protective colours for camouflage, or else bury themselves temporarily or permanently in the sand. The most common are those annelids known as marine bristle worms, in particular two polychaetes – the lugworm (*Arenicola marina*) and the honeycomb worm (*Sabellaria alveolata*), regarded by anglers as ideal bait for certain species of fish.

Lugworms ingest large quantities of sand and assimilate a great deal of the organic substances mixed in it. They evacuate the residue as wormcasts on the surface. The lugworm is common in the most shallow waters, both in sand and at the foot of rocks. It burrows into the sand, forming a tube whose walls are strengthened by a mucus given out from the worm's body. When many individuals of *Sabellaria* are crowded together, their tubes form a honeycomb complex over a wide area of the sea bed.

Many molluscs, crustaceans, echinoderms and fish burrow in sand. Among the molluscs that do so are wedge shells, Venus shells (*Venus*), cockles (*Cardium*), and the razor-shell (*Solen siliqua*), all bivalves with siphons that reach up to and slightly above the sand, for drawing in a current of water containing oxygen and food particles. The outgoing current carries away waste matter. The fan mussel, *Pinna*, which may be a foot (30 cm) long and is one of the larger of the bivalves apart from the giant clams of tropical waters, is found in depths down to a hundred feet

(30 metres). Some gastropods are also present, including the pelican's foot, *Aporrhais pespelicani* (also frequently found on coralline bottoms) a species of whelk (*Nassa mutabilis*), and tower shells (*Turitella communis*). Various squids of the genera *Sepiola*, *Sepietta* and *Rondeletiola*, related to the well-known cuttlefish (*Sepia*), are common on these bottoms. These animals swim well, using the fin running along each side of the body and the siphon from which jets of water are emitted, giving a reaction thrust in the reverse direction, forming a kind of jet propulsion. They tend to keep close to the bottom, inclining the front part of the body downwards. To dig into the sand, they rock the body to and fro using the siphon jet to raise a small cloud of sand which falls down again on to the animal with each movement. Arm and tentacles distribute this covering of sediment uniformly over the body, burying it entirely.

Abundant at various depths on sandy and muddy bottoms are relatives of the common octopus, *Eledone cirrosa* and *Eledone moscata*, easily distinguished from an *Octopus* by having a single row of suckers on each arm, instead of a double row.

Among the crustaceans on sandy bottoms are the shore crab (*Carcinus maenas*), the ilia (*Ilia*

The chalky skeleton of Caulerpa prolifera, *a common green alga*

62

rapidly, but most of the time it remains in the sand with little more than its stalked eyes showing, ready to grasp with its pincers any small animal that passes. Several smaller crustaceans also dig into the sand, including the hermit crab (*Diogenes pugilator*) which practises this habit despite being protected by the shell of a gastropod. It should be remembered that for some animals this burying in sand is a means of ambush, for others a means of self-defence, and for some it may be both.

Other inhabitants of the sandy beds are sea urchins (*Echinocardium cordatum*, *Spatangus purpureus*, *Brissus unicolor*), starfishes, brittle-stars and sea cucumbers. Fishes also dig into the sand, and none has developed this art more than the Pleuronectidae or flatfishes. These have a flattened compressed body and lie permanently on one side. The side in question is almost invariably devoid of colour and, less commonly, so is the pectoral fin. When first hatched a flatfish is the normal shape. As it grows it becomes flattened laterally, turns on one side, and the eye of one side migrates through the skull so that both eyes come to be side-by-side, looking upwards. The mouth is more or less displaced and distorted, so that it tends to open more towards the upper surface than towards the lower or blind face.

The adaptation of the flatfish to its benthonic life is striking. Since it lies on one side, the fish has no need of a protective colouring on that side, nor does it need to use the pectoral fin and the sensory lateral line of that side. Nor does it have to sacrifice an eye to the sea bottom. In swallowing its prey, however, it encounters more difficulty. It has to distort its mouth more or less laterally, so as to avoid ingesting sand or mud.

Above: A close-up of Posidonia, *harbouring numerous epiphytic plants and animals*

Below left: The European shore crab Carcinus maenas. *Below right: Flatfish,* Symphurus ligulatus *and* S. nigrescens

nucleus), and the burrowing crab (*Calappa granulata*). The burrowing crab is big, roundish, with a shell studded with longitudinal strips of tubercles and decorated with red spots; its pincers are flattened, and when they are in the resting position they are laid on the front surface of the body, just as human arms are folded. When the animal wishes to move, it lifts on the tips of the legs and runs sideways very

Not all flatfishes have this joggled symmetry, however: some have been transformed to a greater degree than others, and not all flatfishes are equally tied to the bottom. Some live at very modest depths on sand, others on deep mud, while several make considerable journeys vertically and horizontally in the course of a year. Others are sedentary but have an extensive vertical distribution, being caught all the year round both near the coast and at depths of hundreds of feet. Consequently, it is not strictly correct to associate the flatfish only with sand, since many also live on mud.

The pleuronectids lack nothing in swimming ability, and this ability is not wasted although they do spend a large part of their lives on the bottom. Even as they rest on the bottom, they are at all times prepared for swimming. And when they do swim they keep their eyed surface upward and make undulatory movements that pass along the length of the body, the fish soaring like a magic carpet. At times they keep the body rigid and indulge in long smooth glides. When they return to the bottom, they continue these swimming movements but more slowly, disturbing the sand which then settles on them, hiding under a thin deposit of sediment.

Other flattened fishes, such as skates and rays, are flattened dorsal-ventrally; that is, they are depressed from above downwards. These animals have more or less broad, flat, whitish ventral surfaces, and rest upon them in the manner of the great majority of benthonic fishes. In them, the upper surface back is generally less flattened, and is pigmented, even if not always in a way that makes them blend well with the bottom. The eyes are often higher on the head than in a normal fish, and sometimes are on small protuberances.

Thus flattened fish are numerous on the sea bottom. There are representatives of the bony fishes (Pleuronectidae, and also anglerfish) and of the fish with a cartilaginous skeleton (skate and ray).

Besides the flattened fishes, there are many of the more orthodox shape that live on the sandy

expanses. Among the bony fishes are species of *Trachinus*, the weever fishes, which in repose lie buried in the sand or fine gravel, leaving only the head with its large dorsally placed eyes protruding above the surface. These fishes are dangerous, since the spines on the operculum, or gill-cover, and those of the first dorsal fin can inflict severe wounds. Each operculum contains a poison gland, while the rays of the dorsal fin, which the animal keeps erect protruding from the sand when it lies buried, are equipped with poison-secreting cells. The action of the *Trachinus* poison is similar to that of viper's venom. It is a neurotoxin, causing swelling and discoloration of the tissues, respiratory arrest and a drastic reduction in body temperature. In certain localities the young weever fishes migrate inshore in calm seas, particularly in the morning, and represent a real danger to bathers, who may tread on them.

Much prized as a sea food: Solea vulgaris, *the Dover sole, common in European waters*

Undersea meadows

The submarine meadows appear to form a uniform environment, but in reality it varies significantly, for the most part in accordance with the species of *Zostera* present, the accompanying algae, and the depth. However, one flora and one fauna are to be found on the leaves of the *Posidonia*, another on its rhizomes and on its stalks, others again in the sands on which the meadow is based.

Frequent growths on the stalks are red algae

of the genus *Peyssonelia* and green algae of the genus *Udotea*, found mainly where the firm support, lacking in sand itself, is sufficient for their implantation. Other coralline algae, bryozoans, hydroids and other animals of various groups flourish in calm waters at the base of the meadow, where this extends to depths greater than 85 feet (25 metres).

The world of the animals that live permanently fixed on the *Posidonia* is a complex one, and their

The sea bream, sarpa salpa, *live in shallow water among rocks in coralligenous formations*

life depends intimately upon this. Many species that move about freely among the leaves and the stalks depart when the environmental conditions take a turn for the worse, but there are numerous sessile forms, permanently fixed to the plants, which can withstand the most drastic environmental changes. An examination of the blade-like foliage of *Posidonia* cast up along the shore, reveals colonies of the hydroids *Sertularia perpusilla* and *Obelia geniculata*, and bryozoans like *Microporella*, *Electra*, and *Membranipora*, whose colonies have the appearance of delicate networks: nets in the first genus; perforated ribbons in the second; and small plates in the third.

In connection with these organisms, it should be explained that some hydroids and bryozoans normally settle on those leaves which are the least exposed to the light, appearing in such places more frequently and becoming more strongly developed than elsewhere. This is not only because the planktonic larvae when they change to a benthonic life settle where the optimum ambient conditions obtain, but also because the colonies that arise from them develop in the direction in which such conditions obtain.

The meadows, whether considered as a whole

Above: The sea squirt,
Halocynthia papillosa

Right: The sea slug shown here bears branching processes, known as cerata, along its back, in which are stored stinging cells gathered from the coelenterates that it eats. The animal then uses these stinging cells in its own defence

or in terms of their individual plants, are richer in animal life on the side which is least exposed to sunlight and to the movement of water. The same is true of mountain forests on dry land. The individual trunks of trees are much richer in fungi, lichens and mosses on their northern side, which receives less sunlight.

Nevertheless, not all animals remain on one side of the meadow. Some of those that swim or creep move about on the foliage only at night, while others do so during the day but only when the sky is cloudy, or when the meadow is in shadow. Other animals may inhabit the *Posidonia* in summer but not in winter, behaviour that is linked with the annual cycle of the vegetation, not so much that of the *Posidonia* itself as of the epiphytic algae that grow upon it. These algae suddenly appear in a quite spectacular manner, generally in March in European seas, and develop rapidly. They confer a variable chromatic shading on the meadow as a whole, the shade varying according to place and time. At all times, however, the shade conferred never resembles that of the meadow in winter, when the *Posidonia* is not masked by algae. In conclusion, a collection of animals from *Posidonia* beds varies according to the season, the time of the day and the depth at which it is made.

Above: A banded pipefish is camouflaged as it swims through a submarine meadow

Right: A hermit crab, almost completely hidden by the cloak anemone wrapped around the mollusc shell which it inhabits

67

Besides the hydroids and bryozoans already mentioned, the leaves of the *Posidonia* bear an abundance of diatoms and Foraminifera, algae and Protozoa on which animals like the very small gastropods of the genus *Rissoa* depend for food. These gastropods, using surface tension, can crawl upside-down below the surface of the water holding on by the foot to the surface film. Another animal that feeds on the leaves is the *Idothea hectica*, an isopod crustacean like a large woodlouse with an elongated flattened body. Its greenish colour resembles that of the leaves, providing it with a remarkable camouflage.

Animals found among epiphytic algae which cover the *Posidonia* are nematodes, annelids and also larvae of chironomid midges (from a family of diptera, or flies) and mites. Typical of the *Posidonia* meadow is a small green shrimp, *Hippolyte prideauxiana*, and at the base of the plants are frequently found the sea orange (*Tethya aurantium*), a round siliceous sponge with a rough surface and an orange-red colour.

Above: The Mediterranean wrasse Crenilabrus ocellatus

Right: The sea thrush Labrus turdus, *another wrasse*

Below: the rainbow wrasse, Coris julis, *a common inshore fish*

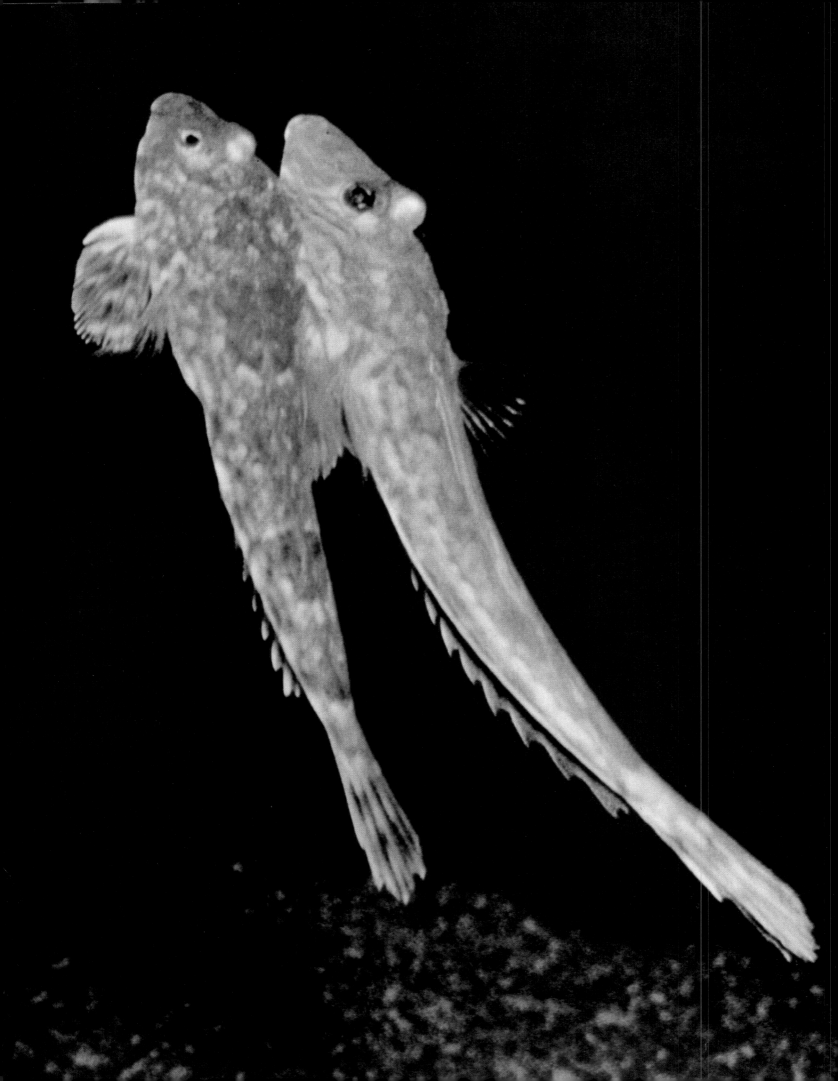

Left: Male and female dragonets swim upwards together in a type of courtship dance

Right: Male and female dragonets, Callionymus lyra; *the male is the larger and has long fins on its back. Below: Dragonets at rest on the bottom of the sea*

The sea orange is by no means peculiar to beds of *Posidonia*, nor does it always look like an orange. It is spherical, often nearly three inches in diameter, and it and a half a dozen related species are found in shallow seas more or less throughout the world, all looking very alike. In the warmer waters it is yellow, orange or red but in colder seas it is more often olive green, and usually smaller, a diameter of about an inch being the more usual.

Its main peculiarity is that it reproduces mainly by budding. Hundreds of specimens have been examined and ova have been found on not more than two occasions. At least 50 per cent of these have, however, had their surfaces covered with buds. These start as warts covering more or less the whole surface of the sponge. Then a stalk appears at the base of each wart as each bud grows out, at first rounded and finally growing into a star-shaped individual at the end of a stalk that has become very thin and long. The stalk ruptures and the bud floats away to settle down and grow into a new sea orange.

Many animals of the submarine meadows are common on other bottoms, for example *Hermione hystrix*, a polychaete worm which is also found on muddy and coralline bottoms, and the sea slug *Chromodoris elegans*, which is also present on reefs. The common squid (*Loligo vulgaris*), swims easily from one part of the bed to another, while echinoderms, such as the feather stars of the genus *Antedon* and sea cucumbers of the genera *Holothuria* and *Cucumaria*, and the hermit crab *Eupagurus prideauxi* are also present.

There is an abundance of fish, particularly bony fishes, and among them are pipefishes, seahorses and wrasses. These are not, however, exclusive to the *Posidonia* beds, since they are also present on the reef and on coralligenous bottoms.

Seahorses look like animated chessmen. They swim erect, driven along by wave-like movements of the dorsal fin. They rest, still erect, anchored by their prehensile tails to plants or other fixed organisms. Pipefishes are extremely long and thin, and are difficult to detect when they remain still, poised vertically among the leaves of the *Posidonia*.

Living in great numbers in the meadows are the wrasses, notably those of the genera *Labrus*, *Crenilabrus*, and *Xyrichthys*. The wrasse family is notable not only for its variability in form and colouring but also for the differences in colour between individuals. In *Crenilabrus scina*, for example, some individuals are red, others green, others bluish grey, and others chestnut grey. This variation in colour has troubled the systematists for a long time. At one point species were based on colours: however, they are now known to vary

The fan shell, Pinna nobilis, *heavily encrusted with a variety of other organisms. This specimen is probably dead, because the shell harbours a fish*

with the emotions, the sex of the individual, the nature of the bottom and other factors, and some classifications have had to be revised.

There are two final points concerned with the submarine meadows that are worth a second thought. One is a purely botanical matter, the other is commercial. As to the first of these there is the question of the pollination of submerged flowering plants. This does not arise in seaweeds which, being algae, do not bear flowers. In *Posidonia* the pollen grains are thread-like and are carried to the female flowers by the currents. In the eel grass, *Zostera*, the pollen is also thread-like. The stigmas of the female flowers are thin and hair-like and are spread out to catch the pollen. Few flowers are so reduced and simplified: in *Zostera* the male flower is reduced to a single anther, while the female flower has only one carpel.

In view of what has been said about the abundance of the submarine meadows, it is natural to ask, in these days when we are being prompted to turn more and more to the harvests of the sea, whether any significant use could be, or has been, made of this vast potential. One positive result of

the dying off of the eel grass was that ducks, geese and swans which fed on the grass were forced to look elsewhere for fodder. More disastrous, in Britain, a whole industry was destroyed. (One of several uses to which the plant was put there was in filling mattresses and the upholstery of furniture.)

No doubt the proliferation of artificial fabrics and materials would have doomed the industry in due course. Yet until a few years ago, at least, there was a factory in Shelbourne County, on the south shore of Nova Scotia, where home insulation was manufactured from eel grass, after suitable drying.

The use of eel grass in Europe for insulation seems to be centuries old. In the Maritime Provinces and in New England, the older houses were insulated with it, and it has many advantages for this. It will not burn and it seems to defy the natural processes of decay, once it is dried, largely because it contains silica, salt and iodine. Its insulating properties are due to the millions of minute gas sacs situated in its tissues, a fact that of course also makes it a good sound-insulator.

Deep-sea mud

Muddy bottoms are generally found at greater depths than sandy bottoms, and gradually succeed them as the continental shelf falls away. They are the most extensive zones of the sea bed, and are of the most varied origin, composition and depth. They extend beyond the limits of the continental shelf, although they frequently begin well above them. Consequently, the muddy bottoms form part of both the neritic province and the oceanic province, and belong as much to the euphotic system as to the oligophotic and aphotic systems. Even if the nature of their sediments were essentially constant, which it is not, the rapid increase of pressure and of darkness with the increase in depth would cause a significant modification of the fauna. As a result, both

sub-coastal forms and those with a wide vertical distribution are found on the muddy bottoms, together with those that are purely deep-sea (bathyphilous) in character.

Muddy bottoms are little if at all affected by the surface currents and the waves, and the waters that cover them are generally calm. Fixed to the bottom with rooting processes are various species of sponge, among them *Thenea muricata* and *Rhizaxinella pyrifera*, sponges with a siliceous skeleton. In the mud live numerous forms of stalked coelenterates, including the dead man's fingers (*Alcyonium digitatum*) and the sea pens (*Pennatula*), which owe their generic name to their shape with its central 'shaft' which carries on opposite sides a series of branches looking like

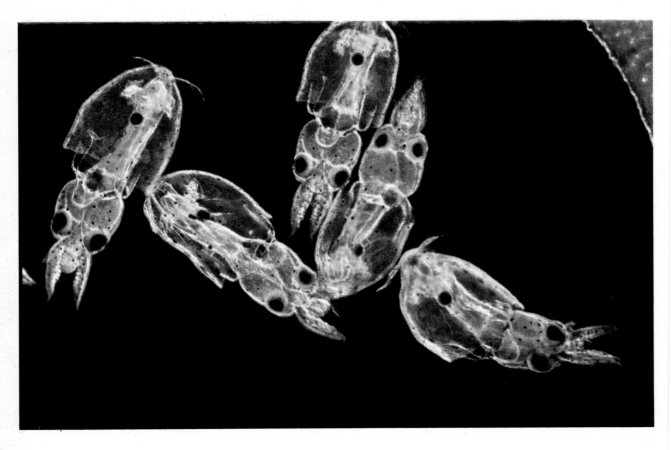

Left: Newly hatched baby squids, Alloteuthis subulata

74

Egg capsules of Alloteuthis subulata *(right), and developing embryos, inside a capsule, of the same animal (far right)*

Right: Octopodidae of the genus Eledone *are commonly found on the muddy bottoms of the Mediterranean*

the vanes of a feather. The handsome *Cerianthus* abounds; it is a large solitary polyp without a skeleton, its column no more than a soft tube, the whole looking like a large anemone with two rings of coloured tentacles, one encircling the mouth, the other on the edge of the disc. Another solitary form is the branching coral *Caryophyllia clavus* which has a remarkable vertical distribution, being found at depths ranging from 100 to 8000 feet (30 to 2500 metres).

On the mud, it is common to find crinoids or sea lilies, like *Leptometra phalangium*, starfishes like the curious *Sphaeriodiscus placenta* and *Anseropoda placenta*, brittlestars like *Ophiura texturata* and the small forms of the genus *Amphiura*, as well as sea cucumbers, such as *Stichopus regalis*. There is an abundance of the polychaete worm known as the sea mouse, *Aphrodite aculeata*, molluscs such as the elephant's tooth (*Dentalium*), gastropods such as *Cassidaria echinophora*, *Dolium galea*, and *Calliostoma zizyphinum*, and bivalves, such as the fan

Cerianthus membranaceus, *a burrowing sea anemone*

Right, top: The ventral or lower surface of a starfish,
Sphaeriodiscus placenta. *Below: The same animal, this time
seen turned the 'right' way up*

mussel *Pinna nobilis* (which is also to be found on
sandy bottoms) and *Isocardia cor*. Also, there are
octopuses of the genus *Eledone*, cuttlefish of the
genus *Rossia*, and squids of the genus *Alloteuthis*.

Crustaceans are common on these bottoms,
among them some of the noted sea foods, like the
mantis shrimp (*Squilla mantis*), the scampi
(*Nephrops norvegicus*), and the so-called red
lobsters, such as *Aristeus antennatus* and *Aristeo-
morpha pholiacea*. Indeed, the muddy and sandy-
muddy bottoms are the source of many of the
most delicious sea foods, since apart from the
crustaceans, these bottoms harbour the mud
mullet (*Mullus barbatus*), the hake (*Merluccius
merluccius*), the codling (*Gadus capelanus*), the
blue whiting (*Micromesistius poutassou*), the
greater forkbeard (*Phycis phycis*) which is a
relative of the hake, and the red bandfish (*Cepola
rubescens*). Together with many species of rays
and guitarfish are found the anglerfish, known
as the frogfish, *Lophius piscatorius*. It is from
Lophius that the 'toad's tail' is obtained, a delicacy
appreciated by all connoisseurs of sea food.

This anglerfish has a body flattened from above
downwards, a wide gaping mouth and long spines
on the mid-line of the head, the front one of
which carries a fleshy flap, used as a lure to bring
smaller fishes within range of the jaws. It is
coloured brown to green-brown with red or
brown mottlings and when freshly caught seems
limp and flaccid. Its whole appearance is unpre-
possessing yet it has now become an important
food-fish. Over 10,000 tons a year are landed in
European waters, mainly from the northern
North Sea and the Bay of Biscay. Most of this is
used in the United Kingdom and Spain. Once its
head is removed and the fish skinned, to take
away its unfortunate appearance, the anglerfish
finds a ready sale, its flesh being very palatable,
and it is enjoyed by many who have never seen
the fish whole.

Tropical coral reefs

The reefs of coral and of madrepores to be discussed in this chapter have an affinity with the coralligenous biocenoses of other seas, but are formations much larger in size. The banks of coral and of madrepora form in these seas like mineral moors around the islands and along continental coasts, sometimes taking the form of real barriers. The most striking example is provided by the Barrier Reef, which extends for more than 1500 miles parallel to the east coast of Australia at an average distance of about 600 feet (180 metres) from the mainland. It is the most imposing work ever constructed by marine organisms.

Coral reefs are ridges formed by the calcareous skeletons of polyps. The coral polyps can live only within limited conditions. They require unpolluted waters with a temperature range of 65° to 96°F. Since they need warm water, corals – and the reefs formed by them – are found mainly in the shallow ocean areas between latitudes 32°N. and 27°S.

Coral and madreporic reefs have not always been confined to so well-defined an area, however. In past geological periods, climatic conditions were widely different from those of today, as has been established by examinations of the different distributions of vegetation in previous ages. It is probable that the first reefs were established in the Triassic seas, which have since been replaced by the land-masses of Germany and the Alpine regions.

The reef-building corals and madrepores never occur at depths greater than about 160 feet (50 metres). When remains of reefs are found at greater depths, it means that the sea bed must have been lowered, in the same way as it must have been raised if similar remains appear above high-tide level.

There are three major types of reef formed by corals: the barrier reef, backed by a lagoon with an island in its waters; the atoll, a reef shaped like an irregular horseshoe or ring, enclosing a lagoon with no island; and the fringing reef, a sea-level flat, near the shore of a continent or an island.

Of these, the atoll is perhaps the most interesting. Because atolls are almost always identical in appearance it was widely believed that they must have had a common origin. One of the first to apply scientific theory to coral reefs was Charles Darwin. It was his belief that all reefs were fringing reefs that grew around islands, and that as the islands gradually sank, so the coral grew both upward and outward at the same rate as the rise in the level of the sea. In this way, the fringing reef became a barrier reef. Finally, the nature of the reef changed again as the island became completely submerged by the sea – leaving an atoll.

Many other theories have been developed to explain the phenomenon of coral reef formation in general, but no single theory of origin accounts for all the facets of coral development. It is known, however, that several factors are involved in the process of their formation, and that to different reefs remarkably different histories are attributed.

Shortly after Darwin had put forward his theory, the great American geologist J. C. Dana found evidence of changes in sea levels in the formation of numerous atolls in the Pacific Ocean. The value of his work is reflected in the present title of the subsidence theory, the Darwin-Dana theory.

One of the more recent theories of atoll formation is that of the Italian geologist, Vinassa de Regny. This presupposes the presence of mountains of the volcanic type, which, on rising up from the sea bed, extend their summits either above the surface of the sea to form an island, or to within a few feet of the surface, to create shallows or submerged reefs. Under both these conditions, corals and madrepores that establish

Palms fringe the shores of a Pacific coral island

themselves on the mountain are able to create an atoll, either with an island at the centre if the mountain is emergent, or one with a basin or lagoon. The coral and madreporic ring never fills out to become a 'disc' because the constructing elements require waters that are clear, well oxygenated and rich in food, and these they find by pressing steadily farther outwards, with a centrifugal expansion, and not by moving into the central part of the atoll, where the water is poor in oxygen and nutrients. It is true that volcanic mountains suffer frequent elevations and subsidences, but if these are sufficiently slow they do not substantially affect the life of these reef-building coelenterates, since they are able to extend themselves along the flanks of the projection, remaining always between the surface of the sea and a depth of about 130 feet.

The colonies that contribute to the building of coral reefs are widely diverse in form: the madrepores, notably the genera *Acropora*, *Porites* and *Seriatopora*, the branching corals, particularly *Meandrina*, *Favia* and *Celoria*, the massive corals, and the solitary corals of the genus *Fungia*, all play a part. Also evident on the reefs are the branching skeletons of millepores, hydrocorallines of the genus *Millepora*, and sea anemones which are often of considerable size and of bright colours, as, for example, in the giant *Discosoma*.

Just as the coralligenous and pre-coralligenous zones of the seas are populated by myriad forms of vegetation and animals, so too are the tropical coral-madrepore formations. Among the varied flora and fauna are found groups of all descriptions: coral polyps, calcareous algae, sessile

Above: A coral-madrepore atoll along the coast of Tahiti

Right: Fiji: the teeming world of a coral reef, exposed at low tide

animals, crawlers, rock-borers, adherers, runners and swimmers, all the principal groups of invertebrates living today, together with innumerable fish.

The predominant invertebrates of the coral formations are the bryozoans, sponges and echinoderms, sea cucumbers, sea urchins, pearl oysters such as *Meleagrina margaritifera*, and the giant clam, *Tridacna*.

The coral-madrepore formations are not all the same in the different oceans. In the Atlantic, for example, there is a rich prevalence of gorgonians, or sea fans, while in the Indian and Pacific Oceans there is a distinct profusion of madrepores. The populations of the different animal and plant groups that play a part in the coral-madrepore biocenosis also vary significantly in their composition, both qualitatively and quantitatively. This should not be surprising in view of the enormous extension in longitude of the formations.

Among the most beautiful fishes of the tropical reefs are the butterfly fishes (Chaetodontidae), characterized by their tall, compact body, the length of their dorsal and anal fins, and the wide range of colours and colour-patterns they exhibit. These animals are excellently adapted to life among the coral reefs, having a protractile

snout which is ideally suited to seizing small invertebrates in the crevices and winding passages in the coral rock.

Another family widely represented on coral reefs are the parrotfishes. These gnaw the coral with their incisor teeth and mince the fragments with the powerful pharyngeal or throat teeth, leaving traces of their passage over the reef in the form of heaps of undigested coral fragments that they have passed through their gut.

Some parrotfishes follow precise itineraries in their daily search for food, and leave behind impressions of their teeth on the formations and also on the accumulations of partly digested coral detritus, like piles of crumbs. Creatures of habit, they seem resistant to any change in the programme of their daily swim, and never vary the place of refuge to which they retreat if frightened.

Among the most beautiful of the parrotfishes are the rainbow parrotfish (*Scarus guacamaia*), and *Sparisoma cretense*, while among the most curious is the bumphead male of *Chlorus gibbus*, in which the adult male bears a large protuberance or bump on the forehead.

The pomacentrids, or clown fishes, are no less interesting than the butterfly fishes and the parrotfishes. Species of *Amphiprion* generally live

Above: Detail of a coral reef, with stag's horn coral to the left, and a giant clam

Left: Three examples of the many different types of coral-madrepore formation

where there are numerous sea anemones, sheltering among their tentacles without being harmed by their powerful venom. Little is known of the relationship between these fish and sea anemones. One theory is that their immunity to the anemones' stinging cells is due to the coat of mucus produced by the fish themselves. This mucus is said to contain substances that inhibit the anemone from shooting out the stinging threads. Recent observations suggest that fish and anemone merely get used to each other.

Among the wrasses, *Coris gaimardi* and *Thalassoma lunare* are frequent visitors to the coral reefs, together with the scorpion fishes, like *Pterois volitans*, and the batfish.

Triggerfishes have an interesting morphology, since when they raise the first long, solid spine of the front dorsal fin, the second spine bends forward and holds the first one firmly in the vertical position. When one of this species is frightened, it hides among the madrepores and corals and raises its spine, and once such safety measures have been applied, the fish cannot be extracted except by breaking the coral. Among the most beautiful triggerfishes are *Balistipus undulatus*, *Balistoides conspicillum*, *Xanthichthys ringens* and *Rhinecanthus aculeatus*.

Finally, the boxfishes, for example *Ostracion*

quadricornis, abound in these localities, together with the porcupinefish, *Diodon hystrix*, and the ballfish, *Tetraodon hispidus*. Both the boxfish and ballfish are considered delicacies in many parts of the Pacific Ocean. Certain precautions have to be taken, however, in preparing these fish for the table, since the adults frequently contain dangerous toxins which may prove lethal if they are not effectively 'cooked out'.

There are several names for these fishes over and above those already mentioned, such as trunk-fishes and puffers. All of these fish are slow swimmers because they lack the abundant muscles of faster fishes. So the amount of 'meat' on them is relatively little. They are usually brightly coloured and since they contain a poison, tetrodotoxin, the colours probably serve as warning colours. Yet despite the small amount of meat they yield, and in spite of the chance they may be poisonous, they are popular food-fishes in several parts of the world. In Japan they are served up as *fugu* and a fugu cook must go to a licensed school to learn how to prepare this dish. The peoples of some of the Pacific Islands are said to roast the fish in its box, and no doubt they know which parts of the fish are likely to be poisonous and should not be eaten.

On the other hand, there is the curious situa-

tion that not all individuals are poisonous, or they vary in the way the poison is distributed in their organs. Moreover, it can happen that box-fishes, for example, may be poisonous in one locality and not poisonous in another, even though they belong to the same species.

The outer surface of the coral formations is very often covered with sands mainly composed of fragments of coral and madrepore, pieces of mollusc shells, echinoderm skeletons and calcareous algae. In addition, at the final stage of the life of the reef, in parts of the formation there are growths of *Posidonia* on the sands, particularly species of *Halophila.*

It is not surprising, therefore, that other animals appearing on the coral and madrepore reefs are more typical of the undersea meadows. In addition, at the bases of the reefs accumula-

tions of detritus of a predominantly organic origin are formed. Among this refuse may live unknown species of animals, since it is likely that many needle fish of the family Syngnathidae have fossorial habits and dig themselves into the sand like certain of the eels. Further support for this view is provided by individuals of the genus *Penetopteryx,* generally pelagic, which have been found on the coral and madrepore formations under a foot or so of fragments. It is clear then, that very little is known of the behaviour of many zoological groups, and that it is not possible to consider the flora and fauna that populate the coral and madrepore reefs in isolation. Many of them share characteristics of forms found in the sands and submarine meadows (and the open waters that surround them). Little, if anything, is however known of the inter-

Right: A blue parrotfish,
Scarus coeruleus, *swims
up through a school of
porkfish* (Anisotremus
virginicus) *in the
American Atlantic*

*Below left: The
trunkfish* Lactoria
cornuta, *or cowfish,
so-called because of the
'horns' that it bears.
Below right: The
batfish,* Platax
orbicularis

Above: A queen
triggerfish, Balistes
vetula, found in
tropical areas of the
Atlantic and Indian
Oceans

Left: The clown
butterfly fish,
Chaetodon
ornatissimus, from the
tropical Pacific

Right: Saddleback butterfly fish, Chaetodon ephippium

Below: Half hidden from sight, the clownfish, Amphiprion *takes shelter in a large sea anemone*

relationships that exist between various types.

For purely zoogeographical reasons it is relevant to consider at this point the sea snakes of the family Hydrophidae, which are widespread in the Indian and Pacific Oceans and neighbouring areas. These animals sometimes swim far out to sea and climb on to the reefs or into the cavities among rocks on small islands, to breed. They exhibit special adaptations to an aquatic life, such as having small scales on the underside like those on the dorsal surface, and a laterally flattened tail which in some species functions as a swimming organ.

A more important adaptation lies in the shape of the body. The front half is usually much more slender than the rear half. To feed they must strike fish with their fangs, in the manner of land snakes. Because they are in a liquid medium they lack the weight and hence the purchase on the substratum that their land relatives enjoy. The heavier rear end supplies the necessary inertia for the slender front end to overcome water resistance. They miss, however, the ability to vary their specific weight by taking in air or by expelling it from the air sac. Sea snakes are generally less than three feet (one metre) long, so they have little in common with the legendary sea serpent. They are highly dangerous, however,

Another clownfish, the vividly marked Amphiprion percula, is one of several species that shelter among the tentacles of large tropical sea anemones

Sponges and sea fans on tropical coral rock

Fossil madrepore coral on the coast of the Red Sea (left), and organic detritus from a coral-madrepore formation on the Mahini Atoll, Polynesia (right)

having grooved fangs and poison glands similar to those of terrestrial snakes. Their venom can be lethal to man but it is rarely used, the snakes generally preferring to flee from danger rather than to attack.

The distribution of sea snakes is curiously circumscribed. It is not surprising that they should keep to surface waters, being lung-breathers, nor that they should tend to hug the coasts. That they should not be more widespread in tropical seas is remarkable especially as one species, the yellow-bellied sea snake, *Pelamis platurus*, turns up on the Pacific coasts of Central and South America in fair numbers and it has even been recorded in Posieta Bay, Siberia, probably carried there in the warm Kuro-sivo current that flows northwards past Japan.

There are two non-marine species. *Laticauda crockeri* lives in a brackish lake on Pennell Islands, one of the Solomons group. *Hydrophis semperi* lives in a freshwater lake on Luzon Island, in the Philippines. The supposition is that

its ancestors became trapped there during a geological upheaval.

Reports suggest that sea snakes sometimes congregate in large numbers, possibly for breeding and at times they are taken in fishermen's nets by the hundred. Even without these aggregations they must clearly be a hazard to fishermen and bathers, for although they are not aggressive they will, like any other snake, strike in self-defence if handled, or on coming into contact with the human body. The numbers of casualties they cause is quite unknown because the regions in which they live do not have the facilities for documentation of this kind. Nevertheless, there are hospital records of patients succumbing to their bite, which is not felt at the time but subsequently produces a paralysis, as well as the tell-tale pair of punctures at the site of the bite.

According to Dr Bruce W. Halstead, in his book *Dangerous Marine Animals*, one species of sea snake has a venom fifty times as potent as that of a cobra.

The open seas

Plankton, already mentioned in an earlier chapter, is a complex of organisms which, whether having the ability of self-movement or not, are unable to oppose the movements of the sea currents, and consequently are transported with them.

This relatively modern definition differs from Issel's (1918) interpretation, which was stated as follows: 'Today the word plankton, created to designate as a whole the flora and the fauna characteristic of the pelagic environment, is coming into general use, but not everyone uses it with this meaning. The custom of calling plankton only those very minute organisms and not to include with it the more considerable species is prevalent. I do not see the reason for such an exclusion, it seeming logical to me to ascribe to plankton the microscopic species and the large cetaceans (whalebone whales, sperm whales, etc) which in the present geological era break the record for animal stature likewise'. While Issel objected to any flora and fauna being excluded from the plankton on the basis of size, he could hardly deny that the flora and fauna of the pelagic division differed widely in their powers of movement. It is on this quality that the modern classification is based: organisms capable of moving freely in the waters independently of its movements form one category – the nekton; and organisms incapable of such movement form another called the plankton. The nekton consists exclusively of animals, while the plankton comprises both animals and plants.

Planktonic organisms of a vegetable nature constitute the phytoplankton, and those of an animal nature the zooplankton. Even this seemingly obvious division is not accepted by all authorities. Some botanists postulate that only plants containing chlorophyll, that is, those capable of photosynthesis, should be included in the phytoplankton. (For in rare cases some phytoplanktonic organisms lose their chlorophyll under particular environmental conditions and thus need to take in from the exterior organic substances which they are not in a position to manufacture for themselves.) It follows, then, that the phytoplankton is essentially limited to the depths that are within reach of sunlight. The zooplankton, however, can descend to much greater depths.

Plankton can be classified with respect to other criteria of a biological, topographical or dimensional nature. In terms of its biology, plankton can be divided into two classes: permanent and temporary. The former includes organisms that belong to the plankton for the whole of their life-cycle, while the temporary plankton includes those organisms that are planktonic only during certain stages of their life-cycle and which for the rest of their life are either benthonic or nektonic.

Plankton is also divisible into neritic and oceanic, the first belonging to the neritic province, the shallow seas over the continental shelf, and the oceanic being characteristic of the oceans beyond. Neritic plankton, like the benthos, is particularly rich but is made up of a proportionately greater number of temporarily planktonic species than the ocean plankton, in which the permanent planktonic organisms are abundant.

Plankton can be divided into six classes: megaplankton, with forms of exceptionally large dimensions; macroplankton, including species that, while relatively small, exceed five millimetres across; mesoplankton, between 1 and 5 millimetres; microplankton, between 50 microns and 1 millimetre; nanoplankton, from 5 to 50 microns; and ultraplankton, less than 5 microns across. (A micron measures one thousandth of a millimetre, and there are 25·4 millimetres to the inch.)

The plants constituting the great mass of phytoplankton are unicellular, and are typified by the diatoms or siliceous algae, the dinoflagellates or peridinians, the coccolithophores with a calcareous skeleton, and the silicoflagellates, with an internal siliceous skeleton. The blue-green algae only rarely constitute a part of the phytoplankton, and only then in the tropical seas. In certain oceanic regions such as the Sargasso Sea, multicellular algae or seaweeds, among them the sargassum weed which strictly belongs to the benthos, drift passively at the surface, constituting a significant proportion of the flora.

The density of the phytoplankton in marine waters varies appreciably according to environmental conditions, such as the richness or otherwise of nutrient salts, and with the season in the same part of the sea. The maximum densities recorded are 40 diatoms per cubic millimetre of water and of 2400 cells per cubic millimetre for the nanoplankton, a record which includes organisms so small that they pass through the meshes of the finest plankton nets. Generally speaking, however, the maximum densities are well below these figures: 0.5 cells per cubic millimetre for the diatoms and 2.5 cells for the nanoplankton.

There is a notable scarcity of planktonic forms in the Mediterranean in general, the recorded values being three to ten times lower than those found under similar conditions in the Atlantic. Sometimes, however, an outstanding quantitative deficiency of phytoplankton is accompanied by an astonishingly wide variety of forms.

The poverty in plankton has repercussions upon the whole of the natural economy of the Mediterranean, since plankton represents the first link of the numerous food chains of marine animals. This scarcity may be attributed to the Mediterranean's general dearth of nutrient salts, and to the low concentrations of nitrogen and phosphorus compounds.

Within the phytoplanktonic group, an equilibrium is often set up between the various species, not only by reason of determined values of the environmental factors, or by competition – a factor in all communities and associations whether animal or plant or both – but also through the production of particular secretions from the individual organisms that inhibit or stimulate growth of neighbouring organisms.

One of the principal laws of planktonic life is to remain within the depth limits of the zone in which environmental conditions favourable to them are found. In relation to such requirements the morphological and physiological adaptations

In their shape, and in the way they slip easily through the water, sharks resemble torpedoes, and they can often be as deadly

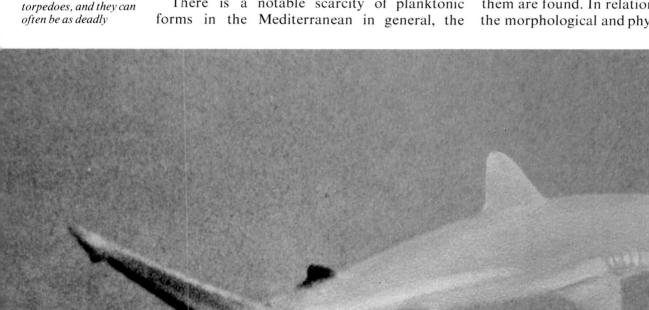

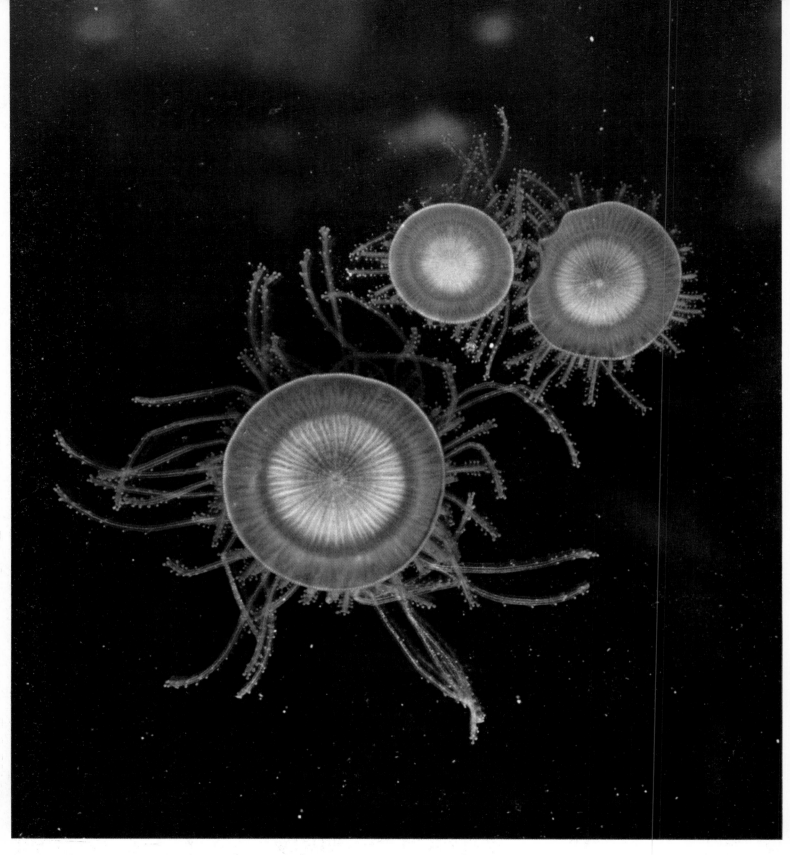

to the environment are fundamentally the same in the plants as in the animals of the plankton. In some diatoms, for instance, the specific weight is reduced by the presence of a large vacoule or by the production of fatty substances, whereas in other diatoms buoyancy is increased by friction with the water produced by means of 'hairs' or spines or by means of a prevalent conformity to the horizontal plane. Other elements of the phytoplankton, for example the *Ornithocercus*, are equipped with extensions that act like parachutes.

Seasonal variations of the temperature of the water, and thus also its density and viscosity, are balanced by compensatory variations in the assorted flotation apparatus of the phytoplanktonic organisms. These variations are, in fact, provided by the adaptations of diatoms – the vacuoles and the fatty secretions which regulate the specific weight. There may also be a variation

Porpita mediterranea, *a siphanophore consisting of a colony of polyps under a float*

94

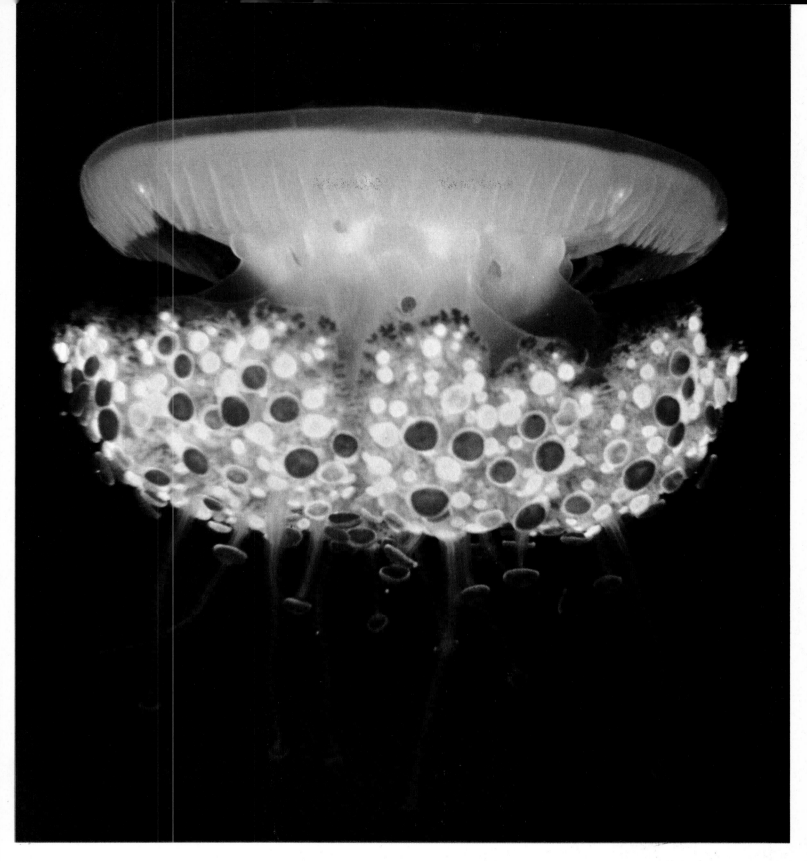

in the length of the appendages, as in *Ceratium trichoceros.*

Cyanophytes or schizophytes, sometimes incorrectly called blue-green algae, are also present in the phytoplankton, but the true algae known as diatoms are of much greater importance. They possess a siliceous shell known as the frustule, made up of two sheaths, one of which fits into the other like a pill-box with its lid. This tiny frustule is very varied in form and is often exquisitely elegant, finely sculptured, almost like an engraving. These algae are mainly solitary, only rarely forming colonies. The most familiar of the true algae, or diatoms, are the species of *Coscinodiscus*, which have a cylindrical shape; those of *Chaetoceras*, which form chains of rectangular frustules bearing filiform appendages; and those of *Thalassiothrix* which are long and extremely slender.

Many dinoflagellates are also abundant in the

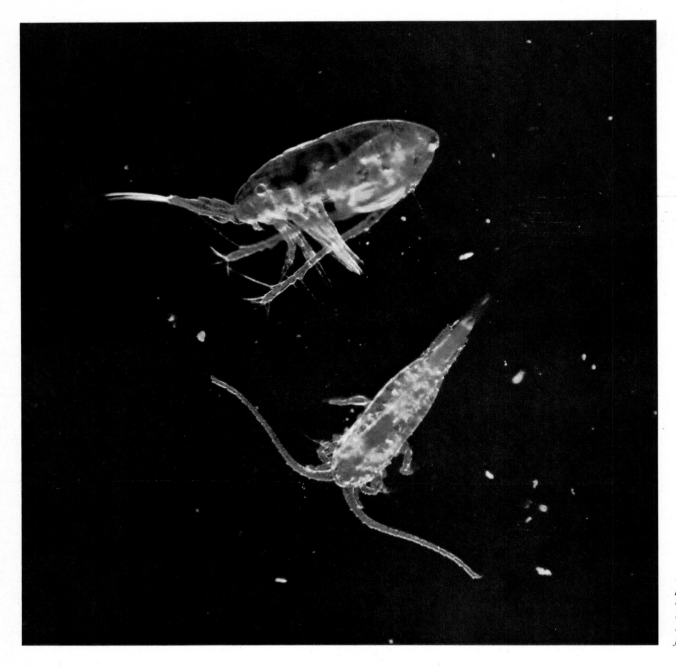

sea. These are unicellular, sometimes colonial, possess chlorophyll and are protected by a cellulose cuirass of a variable form; they have whip-like processes, or flagella, which by their beating carry the dinoflagellate along, giving it an animal-like appearance.

An interesting unicellular alga of the dinoflagellate group is *Noctiluca scintillans* (the scintillating nightlight), a species once considered to be animal, a member of the Protozoa. This derives its name from its luminescence, which displays itself in grand fashion at night, rendering large expanses of sea luminescent, or, as it is popularly called, phosphorescent. *Noctiluca* reproduces with great rapidity by spore-formation and at times hundreds of millions of them may be contained in one cubic metre of water (about two hundred gallons).

Green algae, found in abundance in the coastal benthos, are not common in the plankton. Various forms are visible to the naked eye, the most striking being *Halosphaera viridis*, a species found in the warm and temperate waters of the Atlantic and in the Mediterranean. Each plant is a green sphere approximately one millimetre in diameter.

The bodies of planktonic animals often contain a high percentage of water, which makes them light in weight, and in some cases almost transparent in appearance: some of the jellyfishes, for example, are 96 to 98 per cent water. To whatever zoological group the planktonic animals belong, the heavy parts of their body tend to become reduced, to disappear altogether or to be transformed. *Cymbulia peroni*, a gastropod mollusc, illustrates this well. Its shell is so thin as to

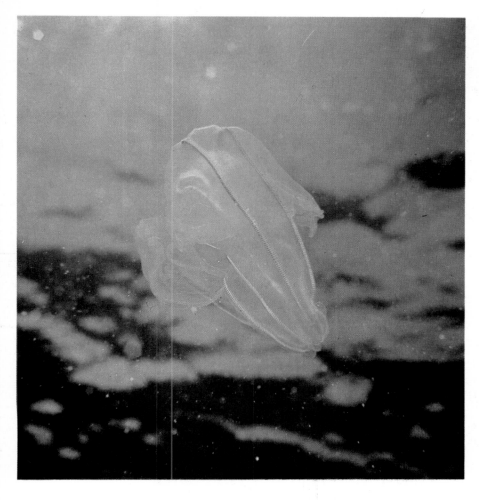

be almost transparent, of a gelatinous appearance and extremely light, very different from the gastropod shells of the familiar sea snails. In some places it is easy, especially in spring, to find shells of these animals scattered on the sea shore, shaped like a slipper. They are so extremely thin and light that it would be possible to suppose them to be the remains of a jellyfish, certainly not a shell.

Protozoa are numerous among the plankton, especially the Foraminifera and the Radiolaria. Common among the former is *Globigerina bulloides*, which is widespread from the surface down to a depth of approximately 3900 feet (1200 metres).

The planktonic coelenterates such as the medusae, are numerous. *Rhizostoma pulmo*, a jellyfish, measures up to a yard in diameter, and is easily recognizable because of its strongly depressed white umbrella bordered in a beautiful violet-blue colour. There are often small fishes under its umbrella, such as the young of the genus *Trachurus*. These young fishes leave their shelter to find food, but return quickly at the first sign of danger. In the winter months *Cotylorhiza tuberculata*, with a flattened umbrella and chestnut colour, is common in many parts of the Mediterranean.

Outstanding among the medusae are the luminescent *Pelagia noctiluca*, found in many areas of the Mediterranean, and the cosmopolitan *Aurelia aurita*, which may be easily distinguished from other species by its saucer-shaped umbrella fringed by short tentacles. A much larger species – and one that may be called a true giant among the medusae – is *Cyanea arctica* of the North Atlantic. Its umbrella attains a diameter of two yards, while its tentacles, dangerous even to man, reach a length of some tens of yards. Other dangerous medusae are those of the genus *Chiropsalmus*, called sea wasps, of the waters of the Philippines and Japan.

All the species cited belong to the Scyphomedusae, some being typical constituents of the neritic plankton, and are thus found in good numbers close to the coasts, while others are characteristic of the pelagic plankton.

The siphonophores are other typically planktonic coelenterates. Their colonial structure has been modified to a free-floating state. At first sight these animals appear to be single individuals, but in reality they are colonies of polyps, beginning with one primary individual, the founder of the colony and the product of a single egg, with numerous secondary individuals that have budded off and developed from the primary individual.

Left: Pelagia noctiluca, *the 'floating night-light', a jellyfish which in shoals lights up the sea at night*

Right: One of the salps, a diaphanous jelly-like member of the plankton, barrel-shaped, even possessing 'hoops'

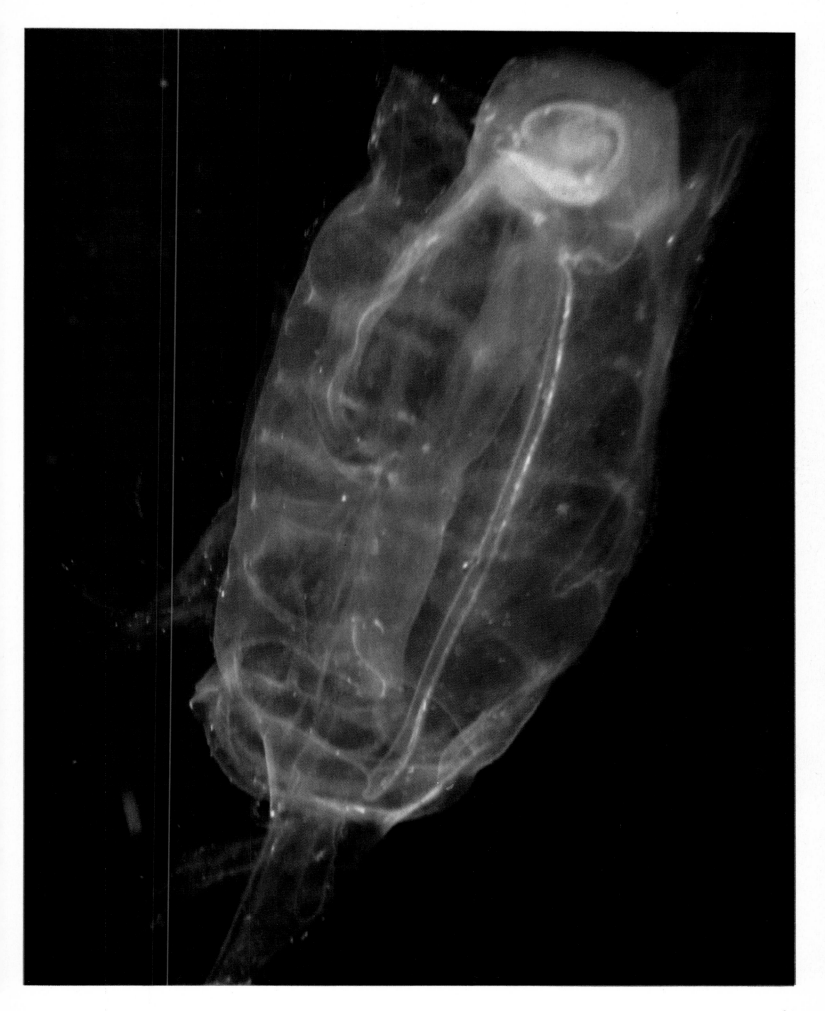

Several types of polyps, differing both in form and in function, are found in one and the same colony. The best known of these coelenterate colonies is the beautiful but dangerous Portuguese man-o'-war (*Physalia physalis*), which is armed with batteries of stinging cells capable of producing painful weals on the body of the human bather, even at times proving fatal. Its popular name derives from its large pneumatophore, or float, surmounted by a crest that bears some resemblance to a small sailing ship. Jack sail-by-the-wind (*Velella velella*), also has the appearance of a sailing boat. It consists of a blue horizontal disc and a vertical triangular crest which functions as a sail. This species, however, is harmless.

Both the Portuguese man-o'-war and Velella live predominantly in warm waters, and are often driven ashore by the prevailing winds and currents.

Above: Shortly after hatching, young green turtles (Chelonia mydas) make their way down to the sea

Various groups of vermiform invertebrates are also represented among the plankton, notably the annelids or ringed worms. Of these, the Tomopteridae lead a pelagic life even as adults and are characterized by conspicuous morphological adaptations to the open waters, such as the transformation of the body appendages into 'paddles'. The larvae of many groups of mesoplanktonic annelids, benthonic as adults, are common in the plankton of the neritic province.

Equally interesting from the biological point of view are the chaetognaths, or arrow-worms. These are little more than transparent torpedo-shaped rods equipped internally with powerful muscles enabling them to dart at small animals, which they capture with two rows of curved 'bristles' or chaetae. The largest, the species *Sagitta maxima*, is only 10 centimetres long, whereas most chaetognaths are only about half this size, or less.

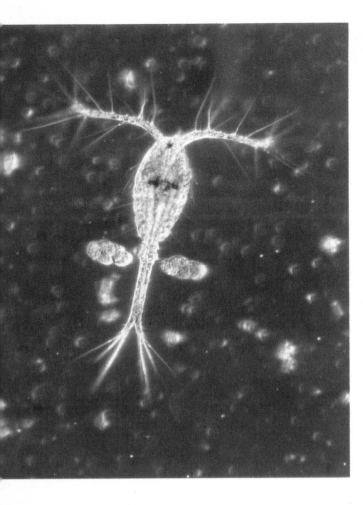

up to 3000 feet or more (about 1000 metres).

A most remarkable example of morphological and physiological adaptation to a pelagic life is provided by the *Pelagothuria natatrix*, a sea cucumber in which the adult is essentially benthonic. *Pelagothuria* resembles a jellyfish in shape. The upturned mouth is crowned by 15 oral tentacles, outside which extend as many long arms joined at the base by a membrane. The cumulative effect is a structure resembling an umbrella. The animal is totally devoid of calcareous plates or gills.

Representatives of the chordates, which are the forerunners of the vertebrates, are also found mainly in the pelagic plankton. These include a number of tunicates belonging to the Thaliacea, such as the Cyclomyaria, the Desmomyaria, and the Pyrosomida.

The Cyclomyaria are barrel-shaped. The giant among them is *Doliolum tritonis*, about five inches long, while the more common *D. muelleri* barely reaches three-quarters of an inch. These animals are transparent, jelly-like and have muscle bands around the body that look very like the hoops of a barrel. They are strongly luminescent at night and are one of the many animals that contribute to the so-called 'phosphorescence' on the surface of the oceans.

The planktonic gastropods belong essentially to two groups, the heteropods and the pteropods. The former have for the most part an elongated body, jelly-like but surprisingly firm to the touch. Its foot is drawn out laterally into two lobes, used as paddles for swimming. Those forms possessing a shell receive only a limited degree of protection from it, since the shell is of small dimensions and therefore does not protect the whole body. Various species of heteropod are widespread down to a depth of over 3000 feet (a thousand metres) and can move actively in these waters with the help of the foot-paddles. The heteropods are generally of relatively large dimensions, the adult forms of certain species of *Pterotrachea* being a foot long (30 centimetres). The pteropods are smaller, with an extremely light and fragile aragonite shell. Members of this group are easily recognizable by their foot which assumes the appearance of wings and function as fins. Species in which this wing function is particularly prominent are sometimes called sea butterflies.

The shape of the pteropod shell varies considerably, ranging from conical (*Creseis acicula*) through tridentate (*Cavolinia tridentata*), to pyramidal (*Cleodora pyramidata*). These small animals are common in European waters and live, like many species of heteropod, in depths of

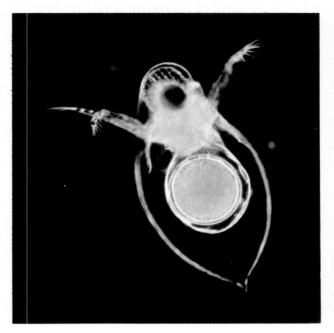
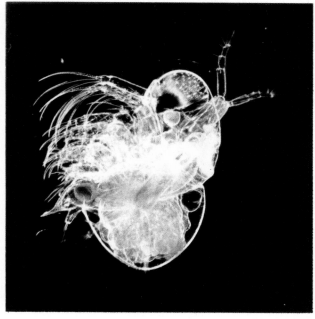

Right: Evadne nordmanni, *a crustacean of the order Cladocera, common in the plankton of the north Atlantic.*
Far right: Podon intermedius, *another cladoceran, living in the plankton with* Evadne

The Desmomyaria or Salpidae include the salps that often accumulate in immense shoals extending over miles of sea in all directions. They sometimes enter coastal waters. *Salpa maxima* the largest, is up to six inches long. These are very similar to *Doliolum.*

The Pyrosomida are extraordinary floating colonies, hollow cylinders whose surface is covered with soft finger-like processes, each of which indicates a separate polyp. The colonies may be over six feet long, and they are found mainly in tropical seas. One species is remarkable for its luminosity and has in fact determined the name of this group, which means fire-body, by the light given out by its symbionts when it is agitated. One of these fire-bodies, four feet in length and 10 inches in diameter, was caught by HMS *Challenger* and a naturalist on board records that he wrote his name with his finger on the surface of the giant Pyrosomida and as a result the name 'came out in a few seconds in letters of fire'.

The Copelata, better known as Larvacea, are also common in the surface plankton. They are tunicates with tails and they look like tadpoles. The luminescent species of this group is *Oikopleura cophocerca.*

Already, in a previous section, we have examined some of the baffling aspects of bioluminescence, or living light, and drawn attention to how little is known of the fundamentals of this phenomenon. There is more to come, for one of the unsolved problems, as spectacular as it is puzzling, has to do with light-phenomena on the ocean surfaces, when the sea lights up all around, and for great distances. Zoologists on ocean-going expeditions have described the light coming on and going out just as if someone had switched it on or off.

Usually, these displays take the form of radial paths of light, radiating out from a centre, like the spokes of a gigantic wheel. Sometimes it is as if the wheel were revolving as successively the beams of light go out and others adjacent to them light up. There has been much discussion about the cause of this phenomenon, but no firm conclusion has been reached.

More easy to understand, and more commonly seen, is the kind of manifestation described by Darwin. He wrote: 'While sailing a little south of the Plata on one very dark night, the sea presented a wonderful and most beautiful spectacle. There was a fresh breeze, and every part of the surface, which during the day is seen as foam, now glowed with a pale light. The vessel drove before her bows two billows of liquid phosphorus, and in her wake she was followed by a milky train. As far as the eye reached the crest of every wave was bright, and the sky above the horizon, from the reflected flare of these livid flames, was not so utterly obscure as over the vault of the heavens.'

Dolphins in a
dolphinarium or
seaquarium.
The animal on the left,
the bottle-nosed
dolphin, is the one most
commonly kept in
captivity

Skeletal support would almost certainly be inadequate and there would be difficulties in temperature regulation. But in the sea the animal finds itself protected and suitably supported by the water. Large size does not adversely affect the capacity of species to live in the sea, a fact amply reflected by the great number of exceptionally large cetaceans, fish and cephalopods found among the nekton.

In addition to their morphology, cetaceans possess a wide range of anatomical and physiological adaptations which equip them for life in the sea. Cetaceans have lungs which give buoyancy and a remarkable capacity for holding their breath for prolonged periods. This capacity stems from the high degree of efficiency with which the cetacean utilizes the oxygen it takes when it breathes. With each breath about 90 per cent of the air is changed, which is far greater than in land mammals. The cetacean receives sufficient oxygen to enrich the blood and to supply this to those parts needing it, leaving less immediately vital muscles in a state of suspended animation. The lungs are not unusually large but the blood can carry an extremely

high concentration of oxygen, and, even when this is exhausted, the animal can remain submerged a little longer by respiring anaerobically (without using oxygen).

All marine cetaceans feed on animals, each group being selective in its choice of diet.

The cetaceans are divided into two sub-orders: the Odontoceti, or toothed whales, and the Mysticeti, or whalebone whales. The Odontoceti have approximately equal teeth and they have no milk teeth; their nostrils form a single half-moon blowhole. They are widespread in all the seas and oceans, with some species inhabiting fresh water. The largest of the odontocetes is the sperm whale, *Physeter catodon*, which attains an overall length of 75 feet (about 23 metres) and preys upon various species of large squid. It is cosmopolitan, but lives principally in warm seas. Another cosmopolitan is the common dolphin, *Delphinus delphis*, far smaller, eight feet (two and a half metres) long, which lives principally on fish.

The mysticetes have rudimentary teeth only during the embryonic stage, after which these are replaced by whalebone, large horny plates with fringed inner edges, which hang down from the upper jaw. These cetaceans are exclusively marine and predominantly inhabit the cold seas, where they feed on small planktonic organisms, in particular prawn-like crustaceans and pelagic pteropod molluscs known as sea butterflies. The popular name of the giant of the group, the blue whale, is something of a misnomer since it is not a true or right whale (*Balaena*) but a rorqual, genus *Balaenoptera*. Rorquals can be readily distinguished from a right whale by the possession of a dorsal fin, which the right whale lacks.

Another group of animals that run to large size is the cartilaginous fishes. There are many species of shark with an overall length of up to 20 feet, but two species reach even greater lengths. These are the basking shark *Cetorhinus maximus*, up to 42 feet (13 metres) long, and the whale shark, *Rhineodon typus*, which measures up to almost 60 feet (18 metres), and is the largest fish living today. Both are inoffensive to man. They have a terminal mouth with extremely numerous small conical teeth and they feed almost exclusively on plankton. The basking shark is mainly distributed in the northern and southern temperate zones, and is especially numerous in the North Atlantic. The whale shark is present in each ocean, although sightings of it are relatively rare. Another giant shark is the great white shark, of tropical waters, certain species of which

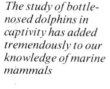

The study of bottle-nosed dolphins in captivity has added tremendously to our knowledge of marine mammals

have reached a length of 40 feet (12 metres).

Giants also exist among the rays, which are cartilaginous fishes related to sharks. Among these some mantas (*Manta, Mobula*) measure 23 feet (7 metres) in breadth and weigh about three or four tons. Despite their size, some mantas are capable of leaping out of the water, when they look like enormous bats, and have received the name of devil-fish. These also are plankton-feeders.

Finally, some colossal oceanic species exist among the invertebrates, namely the giant squids. Their size varies between 40 and 60 feet (13 to 18 metres), including the arms. They are numbered among the prey of the sperm whale, although they do not constitute its basic diet as was once believed. Many cephalopods are preyed upon by various species of toothed whales, but these belong to other families, are more numerous and of smaller size than the giant squids. These huge cephalopods feed on fish and other cephalopods. They are aggressive and will put up a vigorous struggle even with the sperm whale.

Another marine giant, one that has inspired stories of sea monsters, is the leathery turtle or leatherback, *Dermochelys coriacea*, which may reach a length of six feet.

Among 'seafarers' of smaller size, species of 'blue fish' are prominent in the open waters.

These have a metallic greenish blue back and a silvery-white belly. This livery is common among pelagic fishes and provides them with a form of camouflage when they are at or near the surface. To a predator attacking from below and therefore looking up, the ceiling formed by the surface of the sea and the sky beyond appears a brilliant white, which the silvery belly matches. To a predatory bird attacking from above the greenish-blue back merges with the colour of the water below. So the fish can remain undetected unless seen from the side.

Many species of 'blue fish' are of great commercial importance, notably the herring, *Clupea harengus*, the sardine and pilchard *Sardina pilchardus* (the first being the young of the second), and the anchovy, *Engraulis encrasicholus*.

Some species of flying fishes, like *Exocoetus volitans*, dwell in the open sea, along with the pilot fish, *Naucrates ductor*, the common dolphin (the fish, not the mammal) *Coryphaena hippurus*, the rare louvar, *Luvarus imperialis*, the opah, *Lampris regius*, and the ocean sunfish, *Mola mola*.

The ocean sunfish may be as much as 11 feet (about 3½ metres) long, and can be very heavy. Its body is flattened, like a millstone (*mola* being Latin for a millstone), but it swims like a millstone standing on edge, propelled by a tall dorsal fin and an equally large anal fin, both set near the rear part of the body. There is no tail-fin or caudal peduncle, the rear end of the body being rounded.

It has often been noted that sunfishes make no attempt to escape by submerging when harpooned or shot at. They merely give out hideous groans, sounds made by grinding their teeth. Even a healthy sunfish has little to fear because it has a coat of gristle between two and three inches thick under its tough skin – a real bullet-proof waistcoat.

Finally, there are several sea birds that are oceanic, spending most of their time far from land. These come to land only to nest, spending the remainder of their time flying over the sea. The best known are the wandering albatross, *Diomedea exulans*, the storm petrel, *Hydrobates pelagicus*, and to a lesser extent the shearwaters, genus *Puffinus*, and the gannet, *Sula bassana*.

A large tunny is gaffed after being caught

The wandering albatross immortalized by Samuel Coleridge, could equally be classed among the giants, with its wing-span of 18 feet (5 to 6 metres).

There are probably other giants living in the seas whose existence is as yet unsuspected. The whale shark was relatively unknown until this century. The coelacanth, which has become famous as a living fossil, was wholly unknown until 1938. There are sponges as big as barrels, and sea lilies rivalling well-grown saplings. Even the giant squids, on which sperm whales feed daily, are known mainly from a few dead specimens floated ashore. The fact is that deep-sea collecting nets are not the best means of bringing giants to the surface.

The marine abysses

The word 'abyss' and its derivatives are used fairly frequently in natural history literature, often with different meanings. In marine biology, the term 'abyss' is used specifically to denote depths greater than 500 metres (about 1600 feet).

The waters of the abyss are essentially uniform in all the oceans, and the abysses have broad communications with one another. Consequently the abyssal fauna also is generally uniform over an extremely vast geographical range. Further, the abyssal waters are cold, having a temperature approximating to that of the surface temperatures of the polar waters. Their fauna is therefore predominantly similar, if not related, to that of the polar seas. For example, a deep-sea fish may be living at 6500 feet (2000 metres) near the equator but in shallow seas in the arctic.

The ocean bed is covered by an extremely fine sediment, of a variable consistency but generally comparable to that of butter. It is partly continental in origin, due to the transportation of materials brought down to the sea by rivers and dispersed by coastal currents. There may, to a variable extent, be materials of volcanic origin, or even cosmic contributions, but for the main part the sediments are organic.

Innumerable animals and plants in the plankton have skeletons of calcium carbonate or of silica. As they die they sink slowly down, their soft parts decomposing as they go. In some instances the skeletons come to pieces or are eroded by the water as they descend to the bottom. Skeletons of aragonite are dissolved more rapidly than the others, and therefore are the first to disappear. Calcareous skeletons are dissolved more slowly. They can reach the bottom more or less intact at about 16,000 feet (5000 metres). Siliceous skeletons are the slowest to dissolve, and thus can reach greater depths intact.

It follows, then, that the composition of the deep-sea deposits depends in the first instance on depth. The typical pteropod zones, rich in the skeletons of gastropod molluscs known as sea-butterflies that have skeletons of aragonite, are found at depths of less than 3000 feet (about 1000 metres). Further, since the pteropods are transported by the currents, such oozes are found predominantly to correspond with the paths of these currents, in the vicinity of such islands as the Azores, the Canaries, Bermuda, the Marquesas, Cuba and Jamaica, and between Australia and New Zealand. True pteropod oozes are never found at depths greater than 6500 feet (2000 metres).

Globigerina oozes, made up to a large degree of the calcareous skeletons of foraminifera, are found at an average depth of about 12,000 feet (3600 metres), and are rarely found below 13,000 feet (4000 metres). They are more abundant in the Atlantic than in the Pacific, as the former has a lesser average depth, and consequently 'intercepts' the skeletons of the foraminifera earlier in the course of their descent.

The diatomaceous oozes and the radiolarian oozes contain the shells of diatoms and radiolaria, respectively, these making up for the most part between 20 and 70 per cent of their bulk. They occur at a mean depth of around 13,000 feet (4000 metres). They are very rich in clays of mineral origin.

Finally, the deeper abysses are covered in red clay. This deposit, found at depths of more than 18,000 feet (5500 metres), forms up to 25 per cent of the sea-bed in the Atlantic and Indian oceans and 50 per cent in the Pacific.

The red clay is a markedly deficient in calcium, and even the diatomaceous and radiolarian oozes, although they show a modest percentage of calcium carbonate, are significantly poor in this substance.

It is noticeable that animals, both invertebrates and vertebrates, living in the abysses usually have skeletons that are either vestigial or poorly calcified. It could be that their metabolism

is greatly slowed down, by reason of the low temperatures, and that in addition they have difficulty in fixing calcium. The last is due, at least in part, to the sparse quantities of neutral carbonate of calcium present in the abyssal environment, rather than to the high pressures increasing the solubility of the carbonic anhydride in the water, so tending to hold in solution the greater part of the calcium carbonate in the form of bicarbonate, that is of acid carbonate. For example, some deep-sea fishes have skeletons so deficient in calcium that they could be aptly described as rachitic. This description becomes even more acceptable when we consider that calcium is taken from the gut and deposited as bone by the influence of vitamin D formed and active predominantly in the presence of light. The scarcity of the vitamin may be one of the factors responsible for the supposed rickets of so many abyssal fishes.

Furthermore, food is more scarce in the deep waters than in the surface waters, since the food substances that are of interest, directly or indirectly, to animals achieve their maximum production in the sunlit waters where green plants flourish, and where they are to a large extent consumed. Deep-sea animals thus find themselves in quite exceptional environmental circumstances in which they must endure many unfavourable conditions.

However, study of the deep-sea fauna presents some surprises, revealing phenomena that are seemingly contradictory.

Both bony and cartilaginous fishes of the deep seas are generally of small size. They include the frilled shark *Chlamydoselachus anguineus*, unusual in having six gill-openings. This is a comparative giant although its length is only about four or five feet. Yet many species found in the deep waters are giants by any standards. The largest isopod crustacean at present known is *Bathynomus*, the largest brachyuran is *Geryon*, the largest pycnogonid or sea spider is *Colossoendeis*, and the largest sea urchin is *Hygrosoma*; and all these inhabit the abysses. This is less surprising when it is remembered that low temperature delays the onset of sexual maturity, prolongs the period of body growth, and leads to the attainment of a greater size than that reached by related forms in the shallower and warmer waters.

The colour of the abyssal animals is generally uniform, predominantly blackish or reddish, or tending to brown or violet. On the other hand, it is obvious that there is little light in the abysses, and where this is completely lacking, the animals themselves are almost colourless. It is a familiar fact that bioluminescence or 'living light' is a common phenomenon in the depths of the ocean, but its radiations are of a low intensity and do not penetrate far in that environment. Thus the darkness is brightened only by vague glimmers of diverse colours.

This poses the problem we have already discussed: what is the function of the 'lights' produced by many of the abyssal animals? According to some authorities, lights of this nature make it easy for predators to catch their prey, while others see them as recognition signals between the sexes or between members of the same species, where these are gregarious.

Yet one question still remains unanswered: if bioluminescence does serve some such function, of what use is it to a blind fish that is endowed with an individual light? Can it be to focus upon itself the dangerous attention of some predator which is well endowed with eyes?

The lack of bright colours, the blindness and the development of long sensory appendages are obvious features of abyssal animals, but for various reasons the phenomenon of bioluminescence and the remarkable development of the visual organs are not always so easily accounted for.

It has often been affirmed by scientists that the fauna of the deep seas is substantially archaic, composed largely of living fossils. This is partly true, since the adaptations of the animals concerned are for the most part closely related to the modifications in the environment, and the abysses are essentially conservative in this respect, having been less affected than any other environment by the changes that have occurred over the millennia.

It is, however, not possible to generalize, because there are well-known instances of deep-sea forms that are significantly more highly evolved than the neritic forms, notably the deep-sea crustaceans of the genus *Munidopsis*. Moreover, some coastal and shallow water forms of many groups are archaic, examples being the horseshoe or king crabs *Xiphosura* among the arthropods, *Lingula* among the brachiopods, and *Nautilus* among the cephalopods.

In conclusion, the relative uniformity in time and space that is characteristic of the abyssal waters tends of itself to conserve archaic organisms, but the abyssal species are relatively few, and are often distributed over immense areas. Of these, some have originated from the arctic or antarctic coasts, that is, from the high latitudes, and have become deep-sea at intermediate and low latitudes as a result of their narrow temperature tolerance, while others are

Histioteuthis bonelliana, a deep-sea squid that is sometimes captured at the surface

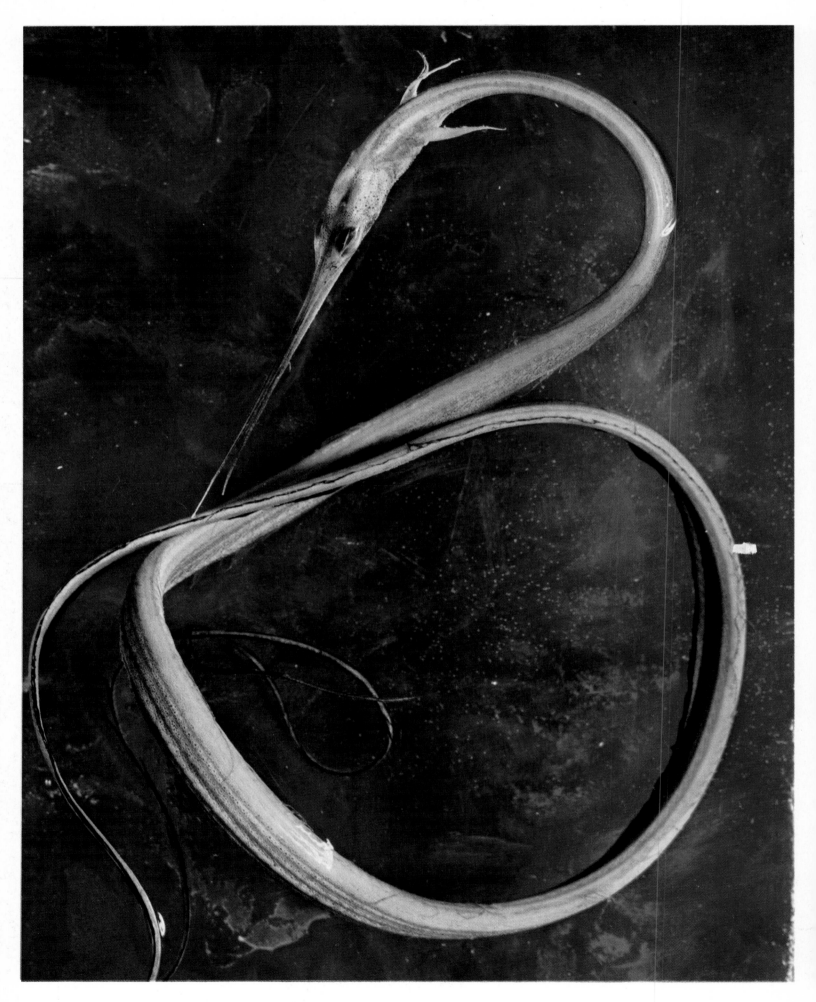

Stomias boa, *a eurybathic fish, widely distributed between about 2,700 to 5,500 feet in all three oceans*

forms that initially had a wide vertical distribution but have survived only in the deepest zones of their former area of distribution. Finally, some forms of coastal origin, but from temperate or warm waters, presumably were thrust bit by bit into the deeper waters through an inability to compete with other species of the warmer waters, or through their lack of resistance to variation in the environmental conditions, or, again, through the greater facility for reproduction in the deeper waters.

Several great zoological groups are represented in the abysses: the Protozoa as radiolarians and as Foraminifera; the Porifera; the coelenterates as the madrepores, antipatharians, and gorgonians; the echinoderms as the crinoids, the ophiuroids or brittlestars, the asteriods or starfishes, the holothuroids or sea cucumbers; and the molluscs, as scaphopods, gastropods, bivalves, and cephalopods.

As the ocean floor at great depths is composed of soft oozes the benthonic animals living on it would run the risk of sinking into it under their own weight were they not supported either by some form of filament (usually a stalk), some form of raft or a rooting process, or by resting on a broad base made up of numerous slender processes that are known as appendices, on which

Nemichthys scolopaceus, *a eurybathic eel, notable for its thin, very elongated jaws*

they can move about without excessive effort.

Nektonic animals also live in the abysses, the most common being cephalopods and fishes. The deep-sea cephalopods frequently have their arms webbed, connected to one another by membranes. One species, *Bathotauma lyromma*, carries its eyes on long stalks. Another species, *Cirrothauma murrayi*, is blind, while the members of the family Histioteuthidae have eyes of different sizes. Some authorities believe that the smaller eye of the histioteuthids serves them in intermediate and surface waters, and the larger one in the abysses. In fact these animals penetrate to a depth of about 20,000 feet (6000 metres), but often at night they come up towards the surface. They should therefore be classified as eurybathic, not exclusive to the deep-sea benthos.

The number, colour and disposition of the light-organs vary no less than the morphology in these bathyphilic and eurybathic cephalopods. Often they are quite bizarre, and the animals possessing them are obviously not all nektonic, some of them being truly benthonic.

The same applies to the fishes, in that among them the bathypelagic forms predominate over the bathybenthonic, both in numbers of species and individuals.

The fishes are principally represented by sharks, rays, chimaeras, herring-like fishes, lantern fishes, eels, rat-tails, squirrel fishes, boarfishes and anglerfishes. They are generally of small size, with a poorly ossified skeleton, and have uniform colours ranging from silvery to black and from red to brown or violet.

Only ten per cent of the species of bony fishes carry light organs but these are much the most common and widespread and are represented by a large number of individuals. Apart from *Ipnops murrayi*, which is blind, and the other forms that are without visual organs or with reduced or non-functional eyes, deep-sea fishes possess eyes not only normally or exceptionally developed by comparison with those of related forms living in the surface waters, but sometimes of an elaborate structure, and known as 'telescopic' eyes.

Several well-known forms are found among the fish that inhabit the deep. These include the hatchet fish *Argyropelecus hemigymnus*, the great swallower *Chauliodus sloanei*, *Stomias boa* (family Stomiatidae), *Cyclothone microdon*, *Vinciguerria attenuata*, and *Gonostoma denudatum* (family Gonostomatidae).

Among the eels, the most interesting is *Nemichthys scolopaceus*, a long fish with a curious head whose jaws resemble the bill of a woodcock.

As with the cephalopods of the family Istioteuthidae, all these fishes are eurybathic; that is, they move to a lesser depth at night, the only difference being that eurythermic species also appear there where and when the temperature of the water is significantly higher than that of the abysses, while the stenothermic species only appear there when the surface water temperature is generally similar to that of the deep waters. This consideration obviously does not apply for those marine areas where any strong vertical or mixed movements of the water transport the animals violently towards the surface (as appears to happen in the Straits of Messina, for example, where deep-water forms are sometimes netted at the surface due to a strong welling up of the water). In the latter case, it is clear that the phenomenon is not solely biological, but also due to water movements.

The wide variety of bathyphilic benthos and nekton tends to obscure the fact that plankton are also represented, albeit on a smaller scale, in the abysses. The principal representatives are the elegant medusae of the genus *Periphylla*, which are widespread in all the oceans.

The Greek philosophers argued that the depths of the oceans must be about the same as the heights of mountains on land, and modern oceanographical research has shown this to be substantially correct. However, Count Luigi Marsigli, writing in 1725, pointed out that for centuries fishermen, and others who should have known better, held the idea that the ocean was a bottomless abyss.

Soundings with a weighted line, for navigation among shoals, with one of the crew in the bows calling the depth, is at least as old as the Ancient Egyptians, as shown by a painting from the tomb of Menna, dated 1422 BC. So the soundings made by Captain James Cook, in the eighteenth century, around the margins of continents, are remarkable more for the methodical and systematic way they were carried out than for any novelty of method. Captain Constantine John Phipps (later Lord Mulgrave) was the first to improve on this by sounding the deep waters. On board HMS *Racehorse*, on a voyage towards the North Pole in 1773, a sounding of 676 fathoms was taken.

During the next hundred years several ships of the Royal Navy carried out surveys in other parts of the world. In the course of these animals were brought up from great depths, either in the dredge or on the sounding lead, and in 1817 and 1818 Sir James Ross, searching for the North-West Passage, with his two ships, HMS *Isabella* and *Alexander*, caused to be made what he described as 'a deep-sea clamm', a grab for taking samples of the sea-bottom. In this, starfish and other deep-sea animals were brought up from as deep as 1000 fathoms. But in those days communications were not what they are today. Apparently as late as 1843 Edward Forbes was still unaware of these events and was able to put forward the idea that no plants grew at depths greater than 100 fathoms, and that at this depth animal life was becoming rare. He was convinced that at depths greater than 300 fathoms the sea was a desolate waste, devoid of life, which he named the Azoic or Lifeless Zone.

Private vessels were also used, like the yacht *Norna*, with the naturalist Savile Kent on board, which in 1870 carried out dredgings in the deep water off Spain and Portugal. Nor was Britain alone in mounting expeditions, although she played a leading role. Yet in spite of what should have been regarded as overwhelming evidence, many scientists stubbornly refused to be convinced that Forbes was wrong. In 1860, however, a telegraph cable lying between Sardinia and Bône needed repairing. It lay on the sea-bed at 1000 fathoms (about 1830 metres), and when raised was found to be encrusted with thousands of sedentary animals. Finally, in an attempt to settle the matter once and for all, in December 1872, HMS *Challenger*, loaded with scientific

Evermannella balbo, a deep-sea fish with a highly distensible stomach, capable of swallowing a fish larger than itself

Gonostoma denudatum, living at depths of about 1,500 feet, has a line of photophores, or light organs, along each flank

Left: Chauliodus sloanei, *a deep-sea fish with a remarkable mechanism for swallowing, which it achieves by opening its gill chambers and throat. Below: A closer view of its devilish teeth*

gear, and with a party of scientists on board, headed by Professor (later Sir) Charles Wyville Thomson, set out on a cruise round the world, to spend three and a half years exploring the oceans. In this time 362 stations were worked, 68,890 nautical miles were covered, and a tremendous collection of specimens of all kinds, as well as a wealth of data, were brought back.

The venture marked a turning point in the study of the oceans.

One expedition after another soon followed, working on much the same pattern as the *Challenger* expedition, but concentrating on more limited areas of the ocean or more circumscribed programmes. In 1877, the US Coast Survey steamer *Blake* set out to explore the Caribbean and the Gulf of Mexico under Alexander Agassiz. From 1880 to 1883 the two French research ships *Travailleur* and *Talisman* surveyed the deep waters of the Mediterranean and the adjacent eastern Atlantic, comparing the deep-sea animals in the two areas.

Right: Lantern fishes,
Argyropelecus
hemigymnus, *carry
photophores, or light
organs.*
*Below: The same fish
seen from the front,
showing its strange eyes*

Meanwhile in 1882, the USS *Albatross* was commissioned by the US Fish and Fisheries Commission for oceanographical work in the tropical Pacific. This expedition took more deep-sea fishes in one haul than the *Challenger* had collected in her three and a half year cruise.

Perhaps the largest individual contribution until modern times came from the nineteenth-century Prince of Monaco, Albert I. As a young man Albert had served in the Spanish Navy and he brought his skill as a navigator, as well as his fortune and his yacht *Hirondelle*, to bear on this burgeoning study. In 1885 he began a series of cruises in the western Mediterranean and the eastern North Atlantic, concentrating especially on the deep waters around the Azores. The work was continued by two of his later vessels, the *Princesse Alice* and the *Princesse Alice II* and the total area covered by the cruises in the eastern Atlantic ranged from the Equator to Bear Island, the island in the Arctic lying between the northernmost coasts of Norway and Spitzbergen.

In brackish waters

The mixing of sea water with fresh water took place in considerable proportions in past geological times, and occurs to a limited degree today, mainly where large rivers flow into the sea. The process has been governed in space and time by different factors, by the speed of flow of the rivers themselves, by the configuration of the sea bed and by the direction and force of the coastal currents.

Where the sea bed descends steeply to a great depth, or where the coastal currents are strong, fresh water from the land dilutes the sea water for the most part only in the coastal zone and at the surface, sometimes significantly altering its chemical composition, its physical nature and its temperature, but without really constituting a distinct environment. The silt transported in suspension by the continental water-courses, in these conditions, is in most cases widely dispersed, only the larger and heavier of the particles being deposited in the coastal area near to the mouth of the river, the movements of the sea carrying the finer and lighter material and transporting it considerable distances.

Along shallow, shelving coasts, however, where the sea bed slopes gently towards the open sea and where strong coastal currents are absent, the majority of the material transported in suspension is deposited near the mouths of rivers to form bars or banks which often break the surface and delimit a stretch of water. Lagoons, lakes, or coastal pools are created, into which there flows a continuous supply of freshwater from one or other of several sources, and at the same time there is a flow of sea water into them.

The salinity of such a stretch of water varies significantly over any given period of time according to the speed of the river and the nature of the tides. In addition, the part closest to the river mouth, or outflow, is composed of fresher water than is the seaward part. When the river is in flood, the whole stretch of water will tend to become fresher, while, conversely, when the tide is high it will tend to be more salty. These brackish waters are classified on the basis of their salinity. With a salinity of less than 0.5 per cent they are said to be oligosaline, between 0.5 and 1.8 per cent they are mesosaline, and between 1.8 and 3.0 per cent, polysaline.

When a lagoon is calm and the river flows slowly into it, the inflowing waters tend to remain on the surface, forming a top layer of brackish water. The stratification thus created, allied to the fact that the density of the fresh water is, for obvious reasons, lower than the density of the brackish and the salt waters, makes the exchange of the various gases between the atmosphere and the deeper strata difficult, and inhibits the development of animal and plant life in them. Certain groups of bacteria, however, flourish under these conditions.

The brackish environment, even within a single stretch of water, is consequently characterized by wide variations in temperature, density, salinity, oxygen content, hydrogen ion concentration, and many other factors of a chemico-physical nature, as well as by the equally significant variations in the biological factors. The total of the animal and vegetable populations of such an environment is called an admixture biocenosis, and the organisms that constitute it originate to a small extent from the fresh waters and to a larger extent from the sea. Few species can be described as typical of this environment, both its flora and its fauna show a remarkable tolerance of wide variations in environmental factors, particularly salinity.

In order to live, both unicellular and multicellular organisms have to maintain a water content which must not vary beyond well-defined limits. Freshwater organisms generally have to eliminate water to prevent an excessive dilution of their body fluids and tissues, while many

Delta of the River Po, on the north-east coast of Italy

marine organisms must draw in more in order to maintain the concentration of salts in these liquids. Some marine forms differ, however, in that their body fluids are in equilibrium with the salt water.

It follows, then, that the organisms must regulate their own osmotic pressure; some, described as osmolabile, show a certain instability in their osmotic balance; while others, termed osmostable, maintain it unaltered in spite of variations in the environmental salt concentration.

The osmolabile organisms do not even control their own volume, and do not survive severe environmental changes, while those that are osmostable maintain not only a relatively unvaried internal concentration, but also a certain constancy of volume. This, of course, involves physiological adjustment and thus an expenditure of energy, a fact made apparent by the

increased oxygen consumption displayed by the organism when under osmotic stress.

As far as the fish are concerned, migratory species which possess a high degree of tolerance to saline conditions execute an inversion of the osmoregulatory mechanisms as they pass from fresh water to brackish or salt water. Instead of continuing to eliminate water through their gills, they begin to absorb it at the level of the intestine and to eliminate salt through the kidneys and the gills. The reverse process occurs in the same fish as it passes from salt or brackish water to fresh water. Such osmoregulation ensures the organism of a reasonable constancy of the chemical composition of the internal medium – even though this is sometimes more salty, and sometimes less salty, than the environment – which allows it to survive. A similar osmoregulation is beyond the powers of stenohaline forms, and they therefore succumb to changing conditions if they accidentally pass from salt to fresh water, or vice versa. When such forms find themselves in brackish waters, they must return to their own environment within a short space of time or perish.

Formations of the lagoon type are found mainly in enclosed seas, such as the Black Sea, the Baltic and the Adriatic, in Europe. In the

This stretch of water towards the mouth of the River Po is probably brackish, and yet supports a wide variety of plant and animal life

Grey mullet, Mugil *sp., a common inshore and estuarine fish*

Adriatic, the north-western coast between the Isonzo and Reno rivers presents a series of stretches of water, the most prominent being the lagoon of Venice, which contains areas of lagoon that are both 'live', that is, affected by tidal events, and 'dead', that is, little affected by the advance of the sea and directly involved in the flow from the continental watercourses. An example of a dead lagoon is that of Caorle, to the north-west of Venice.

122

The 'live' lagoons are rich in algae and in marine phanerogams (flowering plants). The eel-grass *Zostera marina* predominates in the 'live' zone of the Venice lagoon, but gives way to *Z. nana* in the transition zone to the 'dead' part. The species *Ruppia maritima* and *Zannichellia palustris* flourish in the enclosed brackish basins, while in waters which are almost fresh the genera *Potamogeton* and *Elodea* and other 'water-weeds' abound, submerged in fresh water.

Benthonic species are rather rare in the brackish waters, because the bottom is generally slimy and putrid, but a zoocenosis of a marine origin is found, characterized by a species of bivalve, *Venus gallina*, and polychaete worms of the genus *Owenia*. Various other marine forms also inhabit such waters, including acorn barnacles, the starfish *Asterina gibbosa*, the brittlestar *Amphipholis squamata*, the sea cucumber *Trachythyone elongata*, and various molluscs, including *Nassa mammillata* and *Venerupis aurea*. More typical of lagoons, however, are the cockle *Cardium lamarcki* and another lamellibranch, *Scrobicularia piperata*.

Among crustaceans, the predominant species are the shore crab *Carcinus maenas mediterraneus*, and the hermit crab *Diogenes pugilator*. The shore crab can be seen along coasts on the Mediterranean and is well known to gourmets as a tasty and nutritious sea food. Among other visitors to the brackish waters are the shrimp *Crangon crangon*, which is common also in the sea, and some ascidians, such as *Molgula manhattensis* and *Ciona intestinalis*.

Bony or true fishes are also represented in the zoobenthos of the Venetian lagoon. Among these, the most worthy of mention is the flounder *Platicthys flesus italicus*, a flatfish that exhibits an outstanding adaptation, being able to pass quite rapidly from the sea to brackish, or even to fresh water proper, and vice versa. Flounders are often found well up rivers, often miles from the river mouth.

The majority of lagoon fish, like the sardine *Sardina pilchardus*, and the anchovy *Engraulis encrasicholus*, are nektonic; these enter the lagoon from the sea to take their share of the rich food found there. This immigration is temporary, however, since the animals must return to the sea in order to reproduce. Other forms that share this habit are grey mullet, flounder and eels.

The great crested grebe, Podiceps cristatus, in breeding plumage. After the breeding season it repairs to estuaries

A brackish environment is not only found in estuaries and lagoons. Many brackish pools or lakes are formed through the blocking of inlets by sand moved around by heavy seas.

The Italian Lake of Lésina has a bottom largely covered with phanerogams of the genera *Ruppia* and *Althenia*, interspersed with benthonic algae like the sea lettuces, as well as rhodosperms of the genera *Polysiphonia* and *Ceramium*. The phytoplankton are represented by dinoflagellates of the genera *Prorocentrum* and *Peridinium*, and diatoms.

Abundant among the animals of the brackish waters are amphipod crustaceans of the genus *Gammarus*, lamellibranchiate molluscs such as the cockle *Cardium edule*, and the mussel *Mytilus galloprovincialis*. The marine fish of these waters include grey mullet (the genus *Mugil*), red mullet of the genus *Mullus*, pipefish of the genus *Syngnathus*, and, finally, smelt belonging to the genus *Atherina*.

Numerous birds, of many types, may be counted among the community of the brackish waters, inasmuch as many of the same species that frequent the sea coasts and the freshwaters form part of this environment. The flamingo (*Phoenicopterus ruber antiquorum*) nests on some lagoons and pools, using mud to build a conically shaped nest raised slightly above the level of the water.

Some of the euryhaline organisms that visit or populate lagoons and salt pools are also found in the mixture of the fresh waters and salt that is formed 'upstream' or 'downstream' of the mouths of open estuaries or deltas respectively.

Besides euryhaline organisms, there are others in this mixture that are normally classified as marine or freshwater; in the estuaries of certain big rivers there are freshwater mammals, like the cetacean odontocetes of the family Platanistide, and some species of sirens of the genus *Trichechus*, some of which are principally marine and others freshwater.

The four surviving sea-cows – a fifth species, Steller's sea-cow, was discovered in 1742 and killed off by sealers by 1769 – are one species of dugong (*Dugong dugong*) in the Indian Ocean and three species of manatee. The dugong is entirely marine. Of the three species of manatee, *Trichechus senegalensis* is found off the West African coast between the Senegal River and the Cuanaga River. It is marine but enters sluggish rivers. *T. inunguis* lives in waters of the Orinoco and Amazon drainage areas. *T. manatus* inhabits coastal waters and coastal rivers of the south-eastern United States, to the west of Texas, and is also found to the south, in the West Indies and northern South America.

There are four kinds of river dolphins; the largest, only about eight feet (two and a half metres) long, is the susu or Gangetic dolphin, *Platanista gangetica*, which lives in the Ganges and Indus. The Chinese river dolphin, *Lipotes vexillifer*, lives in the Tung Ting Lake, about 600 miles up the Yangtse Kiang, while the La Plata dolphin lives in the estuary of the River Plate.

The river dolphins have beak-like snouts, the jaws equipped with small teeth, and they have a more or less distinct neck, which is unusual in cetaceans. In the marine species the seven cervical vertebrae are usually fused in one mass, although the outlines of the individual vertebrae can be recognized. Even where this does not happen the vertebrae are, so to speak, squashed together, reducing the distance between the back of the head and body are confluent with no sign externally of a neck. The presence of a neck in the river dolphins is regarded as a primitive character, and linked with this are other features of the skeleton that seem to link the river dolphins with the original land mammals from which whales are assumed to be descended. There are eight pairs of double-headed ribs, the breastbone is well developed, and the skull is not so compressed from front to back, as in typical cetaceans, so that the separate bones can be more easily distinguished than in marine species, in which the sutures are largely obliterated in the adult.

If the river dolphins are primitive there are alternative explanations. The first could be that the ancestors of modern whales, dolphins and porpoises, first became aquatic in fresh waters and graduated from the rivers and estuaries to the sea. The other, more probably, is that the river dolphins entered estuaries and rivers from the sea as a means of avoiding competition.

When we look around the oceans today it seems rather ridiculous to suppose that, in the vast volume of water they represent, there could be such pressure from competition for food and living space as to force some dolphins to take up permanent residence in rivers. Such a picture is, however, delusory.

For centuries man has killed whales. The northern seas used to be teeming with whales, but now they are virtually gone. Then the whalers turned their attention to the Antarctic, where large whales abounded. Now those are in turn so reduced in numbers that there are fears they may become extinct or nearly so.

Because cetaceans spend the greater part of their time below the surface they are hidden from us. Two things help to bring home the view we might get if we were able to spend more time watching under water. The first is that the common dolphin, *Delphinus delphis*, which we see only occasionally and in small numbers, have several times been sighted in aggregations estimated at 10,000.

The second is that there are a number of species of dolphins known from half a dozen or so skeletons, and from carcases washed up on the beaches. Yet we know they must be widespread and probably very numerous. There are indications too in other animal species that estuaries particularly offer an escape from competition.

Sticklebacks are common in enclosed waters, particularly *Gasterosteus aculeatus*, which is found in the Mediterranean and throughout northern Europe, and *Spinachia spinachia*, also of northern Europe, which is equally at home in the sea as it is in the waters of estuaries.

Pollution such as that of the Cornish beaches by oil from the wrecked tanker, the Torrey Canyon, *in 1967, is a problem that is today increasing in importance, because of the widespread destruction of marine life, both plant and animal*

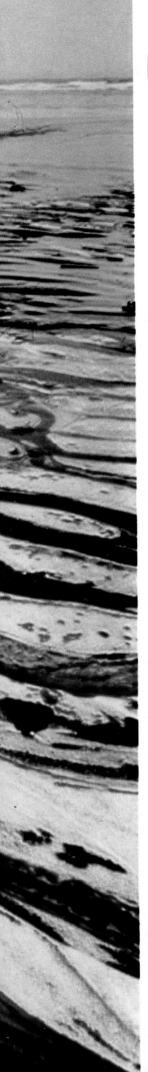

Fears for the future

The science of marine biology originated in Ancient Greece with Aristotle as its principal exponent. The technology of the exploitation of marine resources is much older and more widespread. Marine biology, having flourished in the Grecian epoch, disappeared from view along with many other human achievements under the Roman Empire, and was not resuscitated until the sixteenth century.

Since then it has steadily developed until the present day. Yet now that procedures have been perfected, the explorative and collecting techniques brought to a high level, and the main biological problems satisfactorily brought into focus, there is in fact a considerable risk that the study of life in the sea may have to give way within a few decades to a new discipline: marine toxicology, the study of the poisoning of the oceans.

As industrial development proceeds, waste materials and refuse of all kinds are produced in quantities that neither man nor Nature can safely tolerate. Not only does this waste threaten conservation of the land; the industrial 'boom' also pollutes the atmosphere and the continental waters, and defiles the seas and even the oceans themselves. And it is not only the coastal waters that are becoming progressively more toxic, but the open seas also, and in some cases the ocean depths. As an eminent scientist writes, in many industrial and over-developed zones the rivers are nothing but open sewers: and thus the sea is becoming a sewer 'that never overflows' and one that pollutes even the air.

Yet that warning, like those of other authorities in the fields of conservation and ecology, has only recently penetrated the consciousness of a world which has for too long chosen to ignore or otherwise failed to grasp the true significance of the pollution of the environment. Awareness of the threat is, of course, the essential starting point from which a universal scheme can be launched. Until that vision materializes, the message must be loud and clear: while we continue to abuse the natural resources of earth and water, we increasingly jeopardize the very existence of our own species. The only hope is that men may yet combine to avert such an event.

Bibliography

Barnes, H., *Oceanography and Marine Biology*, Allen & Unwin, London 1959.

Boney, A. D., *A Biology of Marine Algae*, Hutchinson, London 1966.

Burton, M., *The Margins of the Seas*, Frederick Muller, London 1954.

Carrington, R., *A Biography of the Sea*, Chatto & Windus, London 1960.

Darwin, Charles, *Geological Observations on Coral Reefs, Volcanic Islands, and on South America*, Smith & Elden, London 1851.

Deacon, G. E. R., *Oceans*, Paul Hamlyn, London 1962.

Herring, P. J., and Clarke, M. R., *Deep Oceans*, Arthur Barker, London 1971.

Idyll, C. P., *Abyss*, Crowell, New York 1964 and London 1965.

MacGinitie, G. E., and MacGinitie, N., *Natural History of Marine Animals*, McGraw-Hill Book Co., New York 1949.

Marshall, N. B., *Aspects of Deep Sea Biology*, Hutchinson, London 1954.

Ommanney, F. D., *The Ocean*, Oxford University Press, London 1955.

Phillips, C., *The Captive Sea*, Chilton Books, Philadelphia 1964.

Ricketts, E. F., and Calvin, J., *Between Pacific Tides*, Stanford University Press, California 1939.

Riley, J. P., and Skirrow, G., *Chemical Oceanography*, Academic Press, London 1965.

Russell, F. S., and Yonge, C. M., *The Seas*, Warne, London 1963.

Sverdrup, H. V., Johnston, M. W., and Fleming, R. H., *The Oceans, Their Physics, Chemistry and General Biology*, Prentice-Hall, New York 1960.

Thorson, G., *Life in the Sea*, World University Library, Weidenfeld & Nicolson, London 1971.

Wilson, D. P., *Life of the Shore and Shallow Sea*, Nicholson & Watson, London 1937.

Yonge, C. M., *The Sea Shore*, Collins, London 1949.

The photographs in this book have been supplied by:

C. Annunziata: 38 bottom, 39 top, 51, 55 bottom, 93
Archivio B: 77, 110, 112, 114
Beaujard-Cedri-Titus: 30/31
C. Bevilacqua: 21 top, 25
N. Cirani: 3, 4, 5, 9, 14, 16, 23, 119
R. Crocella: 8, 12, 13, 32
P. Curto: 61
D. Faulkner: 24, 41 bottom, 50 top left, 67 top, 83 top,
    83 bottom left, 85 bottom, 88 bottom, 89 top,
    90 bottom, 97 top, 102 bottom
G. Leonardi: 91 right
Maltini-Solaini: 28 bottom, 35, 36, 37 top left,
    37 top right, 39 bottom, 40, 41 top, 44 top left,
    44 top right, 47, 48 top left, 48 bottom, 53 bottom,
    55 top, 57, 59, 64, 65, 66 bottom, 67 bottom, 82 bottom,
    84 top right, 87 top, 88 top, 95, 99
Marka/G. Manti: 108
A. Margiocco: 43 top, 89 bottom, 90 top, 94, 97 bottom,
    100, 101, 113, 116, 117, 124
G. Mazza: 37 bottom, 38 top, 42 bottom, 43 bottom,
    44 bottom, 49, 50 top right, 50 bottom right, 52, 54,
    56 top, 58, 62, 66 top, 68, 69 top, 75 bottom, 76, 86,
    87 bottom left, 87 bottom right, 96, 102 top, 122
P2: 82 top
Photo Researchers: 106, 107
T. Poggio: 7, 10, 11, 18, 80, 81
S. Prato: 19, 34, 78, 91 left
Pubbli-Aer-Foto: 120/121
C. Pulitzer: 48 top right, 50 bottom left, 53 top,
    63 top left, 72, 73, 83 bottom right, 84 top left,
    84 centre left, 84 bottom left
Time-Life/J. Reater: 126
G. Relini: 98
B. Stefani: 123, 125
J. H. Tashjian: 85 top
M. Torchio: 20, 21 bottom, 22, 26, 28 top, 29, 33,
    63 right
D. P. Wilson: 42 top, 45, 46, 63 bottom left, 69 bottom,
    70, 71, 74, 75 top left, 75 top right, 103
World Photo Service: 105